The Mice Who Sing for Sex

The Mice Who Sing for Sex

Lliana Bird &
Dr Jack Lewis

sphere

SPHERE

First published in Great Britain in 2016 by Sphere

1 3 5 7 9 10 8 6 4 2

A CIP catalogue record for this book
is available from the British Library.

ISBN 978-0-7515-6467-9

Typeset in Bembo by M Rules
Printed and bound in Great Britain by
Clays Ltd, St Ives plc

Papers used by Sphere are from well-managed forests
and other responsible sources.

MIX
Paper from
responsible sources
FSC® C104740

Sphere
An imprint of
Little, Brown Book Group
Carmelite House
50 Victoria Embankment
London EC4Y 0DZ

An Hachette UK Company
www.hachette.co.uk

www.littlebrown.co.uk

Contents

*This book is dedicated with love to our incredible parents,
and with a massive hi-five to Richard Boffin*

Introduction

What's the first thing that pops into your head when someone says 'science'? Is it greying men in corduroy flares, painstakingly spelling out the formula for magnesium chloride? Or endless hours spent crouched over the periodic table, breaking out into a cold sweat as you struggle to memorise the elements?

If so, think again. Science can be fun, exciting and even (dare we say it) rock'n'roll. All you need is a little filter to dial down the boring bits and pump up the volume on the more interesting, funny and outright bizarre stuff – like pandas faking pregnancy, how to spot a psychopath or sharks who love heavy metal.

It was with this shared belief that we first set about creating our podcast, and eventually this book – aiming to shed light on the strange, funny and fascinating stuff that you never learned about at school or in the science books.

Throughout the following chapters we will bring you some of the most incredible amazing science stories we've come across, as well as a few of our own personal insights and unique perspectives on the world of science. So who are we? And why have we – a radio DJ and a doctor of neuroscience – decided to venture into this exciting new world of science with a little twist? Let us introduce ourselves properly:

LLIANA

I've always been fascinated by the vast world of science. I suppose it was my dad with his strange yet logical brain leaving endless books on quantum physics lying around the living room, or copies of science magazines for us to peruse whilst we powdered our little noses. My mother, sister and I also shared an obsession for the natural world – collecting bugs, saving injured birds, or visiting endless animal sanctuaries.

But I also had this other side to me, the ying to my yang, one that loved dressing up in strange outfits and going to music festivals to dance to The Flaming Lips through the night. I didn't seem to fit into the mould of a typical science lover (or so I thought).

I headed to university to study Experimental Psychology and it was there that I really discovered how much fun science could be – delving into the deep mechanics of our brains to unlock why we dream, love or even commit crimes.

After university I went off into the world of radio, satisfying my need to explore all sorts of different worlds, meeting and interviewing fascinating people from all walks of life, and yes, even getting to play The Flaming Lips for a living.

JACK

Over twenty years of studying the human brain I found myself becoming increasingly frustrated by the communication barriers that exist between science and the real world. The stereotypical pompous professor, preaching to the masses from their scientific ivory tower, slipping in and out of confounding technical jargon, always struck me as an unlikely vehicle to captivate the imaginations of everyday people with the many wonders of science. This observation inspired me to always try to share my excitement for scientific breakthroughs with other people in a more accessible way; trying my best to speak in plain English at all times.

Many sectors of the media tend to try and 'dumb down' science in an effort to make it more comprehensible to the general public, rather than simply using more straightforward terminology to describe it. I've always found this to be deeply disrespectful to the audience. Worse still, this approach regularly leads to vital parts of the story getting lost in the process and misleading inaccuracies can creep in. All too often the end result is a warped version of just a small part of the story, rather than doing the overall story its due justice. When this happens it genuinely makes me sad, so I have tried really hard not to fall into these traps in my own career. Hopefully I've had some measure of success in this regard, but that's for you to judge!

I strongly believe that if a person doesn't understand the science in an article the blame lays squarely on the shoulders of the storyteller, not the audience. Over the past ten years or so I've seized every opportunity to explain cool science as clearly, compellingly *and* accurately as possible on TV and radio, in books, blogs or speaking engagements. In recent years the desire to do some kind of podcast has always lurked in the back of my mind, but the opportunity never seemed to come my way. Until, that is, I ended up working as science consultant for a TV series Lliana was co-presenting.

LLIANA

I remember the first time I met Jack – it was on the set of a TV show that was I supposed to be presenting, and my initial impression was one of huge relief. The show set out to put psychological phenomena to the test on every day people, and although I had studied Experimental Psychology at university, it had been over ten years since I'd graduated. To say I was a rusty was a bit of an understatement. So here was this young doctor fella – introduced to me as the behind the scenes 'brain expert' – the one who would coach me through and make sure all the actual science bits were totally correct. Phew.

As the week's filming went on I realised that Jack and I had more in common than just a love of psychology. Yes, we were both self-confessed geeks, but we also saw science as dynamic, often hilarious and occasionally ridiculous. The more we talked the more we wondered – surely it wasn't just us who shared this vision? After all Professor Brian Cox was fast becoming the pin up of choice, and 'geek chic' was very much the latest trend.

What if we took the quirkiest and strangest stories from the world of science – those with a real backbone of scientific gravitas, and presented them to the world in a fun and light-hearted way? Could we eventually convert everyone to our mission? Could we convince the world that science was the new rock'n'roll? I quickly invited Jack to join me on my radio show on Xfm for a little feature once a week, bringing the weirdest, most wonderful news from the world of science to my rock'n'roll loving listeners.

We soon found we were genuinely spoiled for choice. From classical music-loving cats to why beer could be good for you, the science world kept throwing up new and interesting research to satisfy the appetites of our audience.

We enjoyed it so much that we decided to delve a bit deeper and turn this little radio feature into a weekly 'Geek Chic's Weird Science' podcast. We found ourselves with listeners from all corners of the planet – as far-flung as China, America and New Zealand. As the podcast grew, so did the support of our loyal listeners, and our belief that science really is something that can appeal to everyone, and anyone. From there, the natural step seemed to be to turn it into a book ...

JACK

The best thing about doing a weekly podcast over so many months is the sense of progress. In the early days we battled over editorial decisions: me fighting for the stories with the greatest scientific validity, Lliana fighting for the conversation-starters, the counterintuitive revelations, the cute and fluffy creatures.

She skilfully eliminated any stories I suggested that fell into the trap of being scientifically-valid-yet-utterly-dull-to-the-non-scientist. In the early months I was shocked by how many of my suggestions fell into this category, I genuinely thought I was down with the kids. But clearly my many years spent in the relative isolation of one neuroscience research institute or another had blurred my perspective. I learned much from these early culls. For my part, I put to the sword anything that was headline-grabbing but didn't have any solid scientific data to back it, or anything where the actual science had been slightly misrepresented in the headline to make the story seem more exciting than it really was. After a few short months we were only suggesting stories to each other that we knew the other would approve of. In other words, we'd quickly figured out how to see both sides of the editorial story selection process. This was tremendously satisfying. Perhaps more importantly, our podcast seemed to go from strength to strength.

We got on like a house on fire right from the outset, but as the weeks and months went by we found that, as our dialogue became more and more fluid, there seemed to be fewer and fewer cock-ups for us to snip out of the podcast at the end. The best bit for me was finally getting to explore the full length, breadth and width of scientific endeavour as a whole, rather than being restricted to the neuroscience literature. There is just so much fascinating research out there about all sorts of creatures, new materials and technologies, outer space, beneath the seas, climate, ecology, food, health and human behaviour that would have otherwise passed me by. And I found that my life was vastly improved with all this extra knowledge. Put it this way, I'm never stuck for interesting things to talk about in a pub, club, festival or London Underground train platform, that's for sure.

That's a bit about us – now on to the book. The next few hundred pages are filled to the brim with some of the most mind-boggling research we discovered over the last three years working on our podcast (including of course, the eponymous mice who sing for sex), as well as plenty of brand new stories we've cherry-picked, which we hope will make you laugh or gasp, or that you'll share with your friends.

You can delve through it in bed, on the tube, or on the toilet – we really don't mind. You can use the stories as great conversation starters in awkward social situations, or simply to break the ice at a bus stop (who doesn't want to know why parrots love The Scissor Sisters?). You could also just use it as a handy guide to show off to your friends when they're having a highfalutin chat (we've dropped the Gravity's Rainbow story more than a few times – it works a treat!).

More than anything we hope you enjoy reading it as much as we've enjoyed writing it. We'll be taking you on a journey through the future, into the minds of psycho-killers and to the depths of outer space. We'll hold your hands as we explore the most bizarre wildlife, the craziest robots, the strangest food, the freakiest sex. Oh, and we'll be finishing off by teaching you how you could make like your favourite superhero.

So strap in, and get ready to have your mind blown by the weird and wonderful world of science.

Love,
Lliana & Jack

1

The Future is Now

In August 1980 two struggling screenwriters sat down and started to imagine what life might have been like had they bumped into their fathers at school. Would they have thought their fathers were cool and wanted to hang out with them? Or would they have thought that they were massive losers and given them a wide berth?

This conversation spawned the script for what would go on to become one of the most popular films of our generation, one of the greatest films of all time. The film would look to the past, taking us back to 1955 to tell the story of a time-travelling young skateboarding dude named Marty McFly who ended up on a mission to get back to the future.

The film's sequel, which came out in 1989, flipped the format and imagined a future where cars flew, teenagers scooted about on hoverboards and people actually made video phone calls (can you imagine?!). The future they envisaged was set in 2015 – 21 October, to be precise – and now, of course, that time has come and gone, giving us a prime opportunity to test the predictions of some of Hollywood's most creative brains and see just how spot on their vision of the future was.

Some of their predictions came true, some came close, some fell by the wayside and some may well have shaped the direction of today's innovations. But the future is now and some of the things that are happening today exceed even the scriptwriters' wildest imaginations.

In this chapter we'll take a look at some of the weirdest, craziest and most astounding technological and futuristic gadgets and gizmos invented, and also take a cheeky glimpse at what our own futures may hold. Marty McFly and Doc Brown, eat your future-loving hearts out.

CHIC FACT: *Back to the Future II* didn't manage to predict the internet, which went on to revolutionise the way we work, play and communicate, but it *did* get a lot of other things right. Video calls, fingerprint payment devices, 'smart' home appliances and wearable tech are all pretty commonplace these days.

Hoverboards

The only way to kick off a chapter about futuristic technology is with the most iconic *Back to the Future II* gadget of all: the hoverboard. And, no, we're not talking about the ubiquitous Swegways that young 'uns bother us with as they cruise up and down the pavements. We're talking about a bona fide, Marty McFly-esque hoverboard – featured in one of the most aspirational, teenage fantasy scenes in filmic history: when our *Back to the Future II* hero 'borrows' a little girl's pink board, rips off the handlebars and then hitches a ride on the back of a car, out of the evil clutches of Biff and his bullies.

It seems that, just like us, many of today's scientists and designers were once tiny little sci-fi-loving kids watching the very same scenes and dreaming that one day they too could be able to hover across a lake without getting their feet wet. Fast-forward twenty-six years from the film's release and the race was well and truly on to see who would bring the hoverboard, with real technology-assisted levitation, to market in time for 21 October 2015, aka 'Future Day'.

Did they manage it? Plenty of companies *claimed* to have cracked it, and *Great Scott!* a few may actually have come close. Even Christopher Lloyd, aka Doc Brown, got in on the hoverboard frenzy when he presented skateboard legend Tony Hawk with a seriously slick-looking hoverboard in 2014. Sadly, that turned out to be an elaborate hoax. (*Et tu*, Doc Brown?) But other authentic attempts have been somewhat more impressive.

Hendo Hoverboard

The Hendo hoverboard was a pretty good attempt and with Tony Hawk steering its development it was bound to be a nifty ride. The team behind it used what looks like a traditional skateboard, but with loads of tiny, spinning, permanent magnets stuck to its underside where wheels would normally be. Others variations of theirs used electromagnets instead, through which the electrical current constantly fluctuates in a very particular way (see Geek Corner). The exact mechanics of how they get them to work are still a bit of a mystery, but whatever their particular secret recipe for magnetic hovering there's a small problem for our Marty-style dreams: you need a magnetic surface such as copper for your Hendo board to repel against, so you'd have to go to special hover rink centres to ride them. Either that or you'd need to lay new roads with a thin layer of copper under the tarmac, which is hardly likely to happen anytime soon. While it could end up being the new Friday night fad – a sort of roller disco for hoverboards – as far as satisfying our childhood dreams of hoverboarding around town with all the freedoms of a skateboarder is concerned, it's a case of close-but-no-cigar.

CHIC FACT: While Hendo's hoverboard attempts didn't quite hit the Marty McFly mark, they ran a great Kickstarter campaign. Their simple developer kits are available so that anyone can tinker about with Hendo's Magnetic Field Architecture™ technology. We've been drooling over one in particular that creates a hovering cube which can be steered using your very own smartphone as a remote control.

ArcaBoard

Arca Space took a different tack, using technology similar to a hovercraft. Instead of magnetic propulsion they used air power, which was created using thirty-six small fans under a rectangular 'ArcaBoard', on which a rider can float above any normal surface. It mimicked some of the earlier and cruder attempts, mainly seen on TV shows, which essentially attached leaf blowers onto the bottom of surfboards. Several powerful jets of air were pointed downwards so that you could skim merrily along on the ArcaBoard a foot above the ground, up to speeds of 12.5 mph. On the one hand it's a great improvement, as it doesn't rely on any special surfaces so technically could travel across a lake; on the other hand, it had better be a small lake, because the battery runs out after six minutes and balancing isn't exactly a breeze.

Omni Hoverboard

Our favourite of the lot, though, is the mighty Omni Hoverboard, another air-powered design self-made by Canadian Catalin Alexandru Duru. This one looks a bit like you're standing on top of two large metal spiders, each one with tiny helicopter propellers for feet, spinning you high up into the air. In May 2015 Duru broke the Guinness World Record for the farthest flight by hoverboard with his carbon-fibre board, hovering 16 metres above Lake Ouareau, Canada. He managed to cover a whopping distance of 275.9 metres – the equivalent of the length of three football pitches, in just ninety seconds. Although he didn't have much choice about the duration of his flight – that's how long the twelve lithium polymer batteries lasted. And we thought our remote running out of juice was a pain in the proverbial . . .

None of the current designs are perfect yet, but we are getting closer and closer to realising our Marty McFly-inspired childhood dreams of what life should hold in the twenty-first century. That said, one thing's for sure – we're sure no one will ever look quite as cool on a hoverboard as Michael J. Fox.

> **GEEK CORNER:** Earnshaw's theorem states that it's physically impossible to get stability in the balance between two opposing magnetic fields, so the Hendo board set-up isn't as simple as having a magnetic field with one orientation in the board and an oppositely aligned magnetic field in the floor. Science is rarely that straightforward. Hendo's creators figured out a way to take advantage of a mind-boggling phenomenon called Lenz's Law to tame the loops of electrical (eddy) currents produced whenever a magnetic field moves over a conductive surface to provide the required stability. Legends!

Hoverboards would be cool for scooting around town but for longer distances a larger, sturdier and faster design would be much better. And if Super Gran can have a car that drives and flies then we want one, too.

Flying Cars

'Where we're going, we don't need roads'
Doc Brown, Back to the Future, *1985*

Sadly, Doc Brown's famous last words as he and Marty rode off towards October 2015 never came true, and our streets are very much still paved with . . . asphalt. Nonetheless, flying cars remain one of the most aspirational pieces of tech from the world of sci-fi, and many of us still dream they may one day become a reality.

> **CHIC FACT:** The earliest mention of the idea of a flying car that we've found was back in 1904. In *The Master of the World*, prescient author Jules Verne wrote of a car that could turn into a plane, boat or even a submarine. Over a century later, we think it's high time that Duck Tours, who zip around London and along the River Thames in their revamped wartime car/boat hybrids, upped their game!

So, How Close are We?

After many underwhelming attempts, it seems we may only have to wait until 2017 – with one particular new flying car design dubbed 'AeroMobil' standing out from the crowd. And if you're wondering what it looks like, just imagine a cross between a car and plane and, yes, you are allowed to call it a 'plar' if you like*. Looking something like a love child of a smart car and a fighter jet, the two-seater operates like any other car. You can park it in normal parking spaces, fill it up at regular petrol stations and shout angrily at other drivers who cut you up as usual. Then, with the press of a button, the wings fold out from the car's sides like an elegant origami bird. *Hey presto!* – you've miraculously converted your ride into a lightweight propeller plane that can take off and land on any grassy or paved surface a few hundred metres long. It's an invention worthy of Q in the James Bond films and, more to the point, it looks pretty damn cool.

> **CHIC FACT:** The first crack at a flying car was in 1928 when Henry Ford designed the 'sky flivver'. Alas, the doomed attempt left one pilot dead and Ford's dream in tatters. However, Henry Ford remained hopeful for the future of flying cars, famously saying in 1940: 'mark my words: a combination airplane and motor car is coming. You may smile, but it will come.'

---

* Geek Chic cannot be held responsible for the cringy looks you may get from your friends and family if you choose to do so. Other side effects may include people not talking to you for up to an hour.

The difference with the AeroMobil is that, unlike some of other *Back to the Future*-loving show-offs flying car attempts out there, the Slovakian company behind it aren't planning on making one just to prove the concept; they actually intend to roll their product out en masse. They've been road and air testing it since October 2014, and videos show the AeroMobil looking as comfortable cruising down a motorway as flying about in the sky, which is promising. Because one thing's for sure: no matter how many other cool innovations and inventions come our way, until we really get to make like *The Jetsons* and whizz about in the sky, sci-fi loving punters like us will keep bringing up that one question on the tip of our tongues: 'when exactly *are* we gonna see this flying car we've been promised since, like, forever?'

We now have an answer to silence us sooner than we thought: 2017.

GEEK CORNER: The AeroMobil flying car is 6 metres long, has a wingspan of 8.2 metres and has a top speed of 200 kph. It has a flying range of around 430 miles, or up to four hours – sufficient to get from London to Aberdeen.

While flying cars are a pretty thrilling prospect, in the meantime we are stuck with flitting across the skies in planes. But the impact of all this rapidly expanding global air travel on the environment is steadily increasing. Is there any way to limit greenhouse gas emissions associated with flying thousands of people through the air every single day . . . ?

Solar-Powered Planes

With air travel rocketing, it makes perfect sense to squint into the future of aeroplanes. Might we one day find ourselves zooming around at the speed of light in our own private pods complete with hot tubs? Or pelting all over the world in futuristic rockets, while robot flight attendants serve us delicious martinis?

> **CHIC FACT:** American Airlines saved $40,000 in 1987 simply by removing one olive from each salad served in first class.

Whatever glamour and luxury future air travel might hold, we first need to find a solution for its whopping great carbon footprint. With the demand for international flights steadily growing, thanks to super-cheap airlines and a rapidly expanding global middle class, the amount of fossil fuels we are burning through air travel is seriously damaging our planet. It's plain to see we need to ensure that soaring demand for flights doesn't kill off our beloved Earth, but then again, surely we can't be expected to give up one of life's greatest pleasures – overseas holidays? Clearly we need a more sustainable alternative to fossil fuels.

One obvious solution is to use solar power, harnessing the Sun's energy and storing it in batteries that can power a plane. The main stumbling block however is that the batteries used to store the excess solar power are usually incredibly heavy, making the energy required just to get their weighty mass up into the air even greater than normal. Batteries for storage are, of course, absolutely essentially because when nightfall comes, or the skies above cloud over, you've

lost all or part of your energy source. And nobody wants to find themselves dropping out of the night sky. The solution? To build a much lighter plane and drastically increase its energy efficiency, which is exactly what Swiss duo André Borschberg and Bertrand Piccard managed to do.

CHIC FACT: During a three-hour flight your body can lose around 1.5 litres of water.

Their creation, the Solar Impulse, is a solar-panelled plane that can fly day and night using energy from the Sun and not a single drop of fossil fuel. It may sound too good to be true, but the dynamic duo decided to put their invention to the ultimate record-breaking test. In March 2015, Borschberg attempted to break a world record for the first round-the-world solar flight using their much-improved second model: Solar Impulse II. With a wingspan of 72 metres*, even bigger than its predecessor, and more than 17,000 solar panels, Solar Impulse II's circumnavigation of the world got off to a pretty good start. It travelled from Abu Dhabi through Asia and across the Pacific Ocean to Hawaii, but at that point heat damage to the batteries eventually put a temporary stop to the intrepid project. Happily, after a few frustrating and expensive months of downtime, in which they had to find extra funding to make repairs, they did finally get going again.

Every solar-powered cloud has a silver lining, however, as Borschberg still managed to set a record for the longest solo (and as it happens, solar) flight, staying up in the air for a whopping 118 hours 52 minutes. He must have got pretty lonely up there all on his own in the tiny little cockpit, not to mention sleepy, as he was only allowed catnaps of no more than twenty minutes at a time throughout. But, like Phileas Fogg, Borschberg refused to be put off and has vowed to complete his solo-solar-round-the-world-trip.

---

* At 72 metres wide and only weighing in at 2,300 kilograms, Solar Impulse II has a wingspan wider than that of a Boeing 747 but with a mass more or less equivalent to that of a Ford Fiesta.

Whatever happens, in the long run it'll be worth all the expense and discomfort as it's already done a fantastic job of raising the profile of solar power as a genuinely viable, clean, green future of air travel.

GEEK CORNER: The price of solar panels has tumbled over the past few years. Deutsche Bank projections estimate that generating electricity using solar panels will match the cost of gas-powered electricity by 2020 in 66 per cent of the world. The International Energy Agency has also predicted that solar power will account for 27 per cent of global electricity generation by 2050, surpassing that contributed by coal-fired power stations.

Flying cars and solar-powered planes are an exciting prospect but in the interim we are still likely to find ourselves stuck in endless traffic jams down here on terra firma. So what innovations do the future of our roads hold?

LEGO Roads

There are few things more frustrating than sitting in a traffic jam, particularly when it's caused by roadworks. You patiently wait for the temporary traffic lights to turn green, only to find, as you finally drive past, that no one is actually working on the cordoned-off stretch of road.

CHIC FACT: The word LEGO comes from the first two letters of the Danish words for 'play well' – Leg Godt!

Abandoned or not, roadworks on our streets and motorways can be nerve-frazzlingly annoying and they always seem to turn up at the most inopportune times: slowing you down when you're running late for an important meeting, class or wedding; sometimes lasting for weeks, months or even years with no sign of anything actually being achieved to make all that disruption worthwhile. Why is that? Well, one part of the problem is that roads are built of asphalt, which is time-consuming and laborious to drill up and re-lay. If only there were a different way . . .

Step forward LEGO! Yup, that's right. As if we didn't already need to thank Denmark enough for one of our nation's favourite toys, it now turns out that LEGO has also come up trumps as a muse for Europe's finest town-planning innovators. A group of Dutch engineers have designed a LEGO-inspired plastic road and, mustering all their creative genius to dream up a breathtakingly original name for this invention, they've ended up calling it 'PlasticRoad'.

CHIC FACT: In several places in the world, the Indian city of Jamshedpur, for example, waste plastic and old tyres are routinely collected, ground down into small particles and added to asphalt. This not only helps to get rid of these unwanted materials without contributing to landfill sites but also makes the roads up to 50 per cent more durable.

The idea was to build the road in sections from prefabricated slabs that are transported wherever they are needed and snapped neatly into place in a manner not dissimilar to LEGO. Different compartments inside the toughened plastic slab contain all sections to contain all the cabling, pipes and other subterranean shenanigans.

Essentially what this means is that roads can be quickly laid overnight, with specific sections being removed and replaced before the rush hour traffic starts again the next day, causing minimal disruption.

CHIC FACT: In the USA plans are afoot to use roads to generate electricity by embedding solar panels into some of the country's 72,000 km^2 of road surface.

Better still, because the PlasticRoads are made entirely out of recycled plastic, side-stepping the toxic emissions associated with laying conventional asphalt, its eco-friendly rep is sky-high. Plus, the end result is exceptionally tough so it'll last longer (see Geek Corner).

PlasticRoads are currently being trialled in the Dutch port city of Rotterdam, but there's a long road ahead ('scuse the pun) before the LEGO-inspired thoroughfares actually hit our streets (sorry, we can't help ourselves). However as 'the future' is more or less infinite, we have no doubt that this new innovation will, in some guise, at some point, become a LEGO-tastic reality. A future free of roadwork-related traffic jams? We'd be very happy to live in such a world.

GEEK CORNER: Unlike conventional tarmac, the PlasticRoads should be able to withstand temperatures of −40 to +80°C. Best of all? They are predicted to last three times longer than conventional roads before needing replacing.

So far we've dreamed about making like Marty McFly with hoverboards and flying cars, and contemplated the future of our roads. High time, then, that we turned our attention to another iconic futuristic invention of Back to the Future II *– self-lacing shoes.*

Self-Lacing Shoes (and other wearable tech)

Remember when you were a little tyke and your mum or dad taught you how to lace up your shoes? *Over, under, around and through; meet Mr Bunny Rabbit, pull and through.* The struggle was real, the lacing was hard, but kids of the future may soon have it easy, because self-lacing shoes are finally a reality.

Unsurprisingly, the person who got to keep the very first pair was none other than our old friend Marty McFly, aka real-life actor Michael J. Fox.

Nike presented Michael with his very own self-lacing shoes on – you guessed it! – 21 October 2015, the day *Back to the Future II* predicted they would have become a part of our everyday lives. Michael demonstrated the uber-cool-looking Nike Air MAGS live on TV and then tweeted a letter from the shoe designer thanking him for making self-lacing shoes popular so many years before they actually existed.

CHIC FACT: The original designer of Marty McFly's Nikes was the aptly-named innovator, Tinker Hatfield. He also led the design of Nike's most famous product: the Air Jordans.

In other *Back to the Future II* fashion news, Marty's infamous self-drying jacket has also been invented: one press of a button and the jacket inflates, drying it to 90 per cent in just under a minute. Yet another example of sci-fi inspiring real-life tech. Today's wearable technology has gone way beyond the world imagined by film-makers. Here are a few examples of this new tech-wear:

GEEK CORNER: A French company named Zhor Tech went one step further than self-lacing shoes: you can tighten or loosen their shoes using a phone app, get feedback on how many calories you've burnt that day *and* adjust the temperature using an in-shoe air conditioning system.

- Leggings that double as a yoga instructor, using in-built movement sensors to give you verbal guidance and feedback on technique as you stretch into each pose.
- The world's first 'smart-bra', which measures heart and breathing rates as you work out, helping you monitor changes in your biometric signals to achieve your athletic goals.
- A hairband that claims to bring back thinning hair with the use of some nifty little lasers (Donald Trump, take note).
- The now-ubiquitous smart watches that can accomplish an astonishing amount of things, from tracking your daily activity to sending emails and making phone calls like a real-life Dick Tracy.

CHIC FACT: In *Back to the Future II* the famous 'self-tying laces' don't actually tie up at all – they simply constrict and tighten; no tricky bows or knots involved whatsoever.

If you're looking to nab yourself a cheeky pair of self-lacing Nike Air MAGS, bear in mind the entire limited edition release collection was auctioned off, with all money going to the Michael J. Fox Foundation for Parkinson's Research. That said, super-fans don't lose hope just yet – last time we checked there was a pair for sale on eBay for a swoon-inducing £5,000.

Fashion and technology will no doubt continue to merge as the years go by, but one of the most exciting examples of this in the eyes of your geeky (and chic!) authors is the one we describe over the next few pages: something that looks hot, incorporates cutting-edge tech and takes us one small but important step closer to cleaning up our planet.

Swimming Costumes That Clean the Seas

It's the ideal scenario: you're somewhere tropical on a well-earned beach holiday, lazing about drinking cocktails in the sun, wading out

for a nice, long, cool dip in the ocean and then patting yourself on the back at the end of the day, safe in the knowledge that you've made a positive contribution to the wellbeing of the planet.

We could all soon have a piece of this dream thanks to engineering wife-and-husband super-duo Mihri and Cengiz Ozkan, who've created a brand new material they call 'Sponge', which mops up dangerous chemicals, cleaning up our oceans. This highly porous material can be incorporated into swimwear, wetsuits or any other material headed for the endless blue, simultaneously repelling water and trapping toxic chemicals within its core in a space-age, sugar-based, molecular sponge.

CHIC FACT: The world's most expensive bikini is worth £20 million and made almost entirely out of 150-carat diamonds and platinum. Presumably anyone wearing that on a swim would sink straight to the bottom – which begs the question, why bother?!

3D printers have already been used to create bikinis made of this super-material, which any sea-loving lady could happily take on her holidays and wear for her daily dip. And thanks to its special structure, the material can hold up to twenty-five times its own weight in contaminants – not bad.

GEEK CORNER: Sponges made from a plastic known as PLGA can be used to 'mop up' dangerous cancer cells, preventing them from reaching new areas and even letting doctors know whether a cancer is spreading.

But hold on a second! Doesn't this mean people will end up swimming around with some seriously toxic substances rubbing against their skin? Fortunately, that's not the case. The absorbed chemicals are, in fact, safely locked away in the inner core of the material, and only released upon exposure to temperatures in excess of 1,000°C – hardly likely to happen accidentally. What's more, the average swimmer could wear their sweet new threads at least twenty times before it needs to be recycled. Cleaning our seas and oceans simply by wearing a special bikini or wetsuit while on holiday? Being an eco-warrior might not be such hard work after all . . .

CHIC FACT: The first bikini as we know it today was brought to market in 1946 with the strapline: 'reveals everything about a girl except her mother's maiden name'. However, it seems that bikinis have been around for some time. Ancient Minoan paintings from 1600 BC and a fourth-century AD Sicilian mosaic – informally called 'the bikini girls' – show women in two-piece swimsuits.

While we're on the topic of technology that enables us to take a dip in the sea and contributes to its wellbeing at the same time, yet more brilliant minds are looking deep beneath the waves and envisaging a future in which we could actually end up living there.

The Future's Bright, the Future's ... Underwater?

From the legend of Atlantis to the world of SpongeBob SquarePants, underwater cities have long captured our imaginations, but now one company claims it will make them a reality.

Shimizu Corporation, based in Tokyo, has been working on the designs of its Ocean Spiral city since 2014 and, we have to admit, it looks pretty damn cool. The underwater city itself, which they hope to move people into by 2030, will be set within what looks like a giant bubble with an entrance 200 metres below sea level – safely below the potential disturbance of whatever swell conditions are going on at the surface, but still just about illuminated by sunlight.

Offices, hotels and research labs are a few of the buildings that could comfortably sit inside the main 500-metre-wide sphere, which has been dubbed the 'Blue Garden'.

The whole thing would be anchored by an 'infra-spiral' – which looks a bit like a giant corkscrew – to a renewable energy-powered factory fixed to the seabed that would provide the electricity, drinking water and oxygen required by the city's inhabitants. Plus, the whole thing could potentially be powered by the CO_2 we breathe out, in combination with other waste products that could be converted to methane gas by micro-organisms. The most exciting part? Any citizen would be able to head to the 'deep-sea gondola lobby' and catch a lift down the infra-spiral to check out what's happening down below for an impromptu deep ocean mini-break.

> **CHIC FACT:** A 1,300-year-old city named Shicheng – roughly translated as 'Lion City' – remains almost fully intact yet submerged beneath the surface of a lake in China. The ancient city was on land until 1959 when all of its 300,000 residents were relocated after the Chinese government's decision to build a dam and flood the area.

The Ocean Spiral city isn't only there to look impressive, though; it actually solves quite a few problems we face up here on dry land. The combination of overpopulation and lack of housing, not to mention limited land space on which to build, are all huge issues – and let's not forget that the ocean covers 70 per cent of the Earth's surface and is pretty damn deep, so the potential benefits associated with building these underwater cities is massive.

The catch? It's estimated it would to cost over £16 billion to build and Lord knows what fresh hell getting planning permission is going to be. We reckon it would all be worth the hassle if the pay-off is the opportunity to sit, daydreaming at your desk and looking out of the window as a hammerhead shark swims lazily by.

CHIC FACT: The myth of Atlantis was first told by Plato in 330 BC in the dialogues *Timaeus* and *Critias*. It was only in 1881 that a writer named Ignatius Donnelly suggested that it might in fact be a real-life lost city.

As many of us may one day find ourselves living under the sea in these under-water cities, it would probably make sense to figure out how to feed ourselves without having to go to the trouble of travelling back up to the surface to collect food. Just imagine having to spend half an hour in a decompression chamber every time you wanted to tend your vegetable patch. But fear not: a solution is at hand, as an intrepid bunch of submariners have been working hard at the challenge of figuring out how to grow plants in greenhouses under the sea.

Underwater Farming

You may have thought that going fruit-picking Down Under on a gap yah seems pretty exotic, but how about picking strawberries deep beneath the sea? This may not quite be ready for any Tom, Dick or Henrietta just yet, but underwater farming may well feature in our not-too-distant future. And while we may have to wait until 2030 to live in an underwater city, an underwater farm has already been successfully trialled.

From 2012 to 2015 scuba-farmers Sergio Gamberini and his son Luca tended their crops in 'Nemo's Garden', which just happened to be 5–10 metres (or 20–30 feet) under the sea, in a little bay in northern Italy.

The crops themselves were grown inside transparent biospheres, which looked a bit like giant jellyfish-shaped balloons attached to the seabed by tentacles made of rope. As they were only ever submerged at a maximum depth of 10 metres they still got plenty of sunlight, and being deep under the sea, away from the changeable weather above the surface, meant they were easily kept at a plant-friendly temperature of around 25°C. Toasty.

Inside these underwater greenhouse bubbles, water condenses on the roof and then drips back down onto the plants below to keep them nicely watered. Plus this process creates the humid environment that many crops thrive in – around 85 per cent, in case you're interested. Better still, there are no problems with crop-munching pests or beasties, so no nasty pesticides are needed. Great news for us fruit-and-veg-munching humans; perhaps even more so for our old friend, the bee*.

The three-year trial in northern Italy was a great success and our heroic scuba-farmers managed to grow an abundance of tasty strawberries, beans and lettuce. They also managed to grow shed-loads of basil, handy for all the Italian mammas making the famous pesto sauce of the local region, Liguria – an area where traditional crop farming can be tricky thanks to rocky terrains, overcrowding and the occasional landslide or flood.

> **GEEK CORNER:** The future of farming isn't just under water; it could also be up in the skies. In South Korea a 'SkyFarm' has been designed to grow crops on a tree-like skyscraper, not only providing food, but also helping to clean the air in the midst of a busy city.

So our Noli Bay scuba-farmers have proved their point: it *is* possible to grow delicious fruit and vegetables underwater in areas where land and climate conditions are far from ideal. Now the trial is over, there's no reason why underwater farms can't be rolled out to other areas around the world.

The biospheres are cheap to run, don't need pesticides, free up precious land and, with no need for thermal regulation, require very

---

* In 2015 a study finally proved that certain pesticides are contributing to the mass death of bees and that without bees – responsible for cross-pollination of a huge proportion of the crops we rely on – we are, quite simply, screwed.

little energy. Who knows? They really could be the farms of the future.

CHIC FACT: The world's first underground hydroponic urban farm opened in London in 2015, in a series of disused Second World War tunnels up to 33 metres beneath the streets. They are located beneath the London Underground's Northern Line in Clapham, and were originally used as air raid shelters to accommodate up to 8,000 people. Fancy a bite of some food grown in such ingeniously repurposed historic settings? Head to London's Borough Market and keep an eye out for fruit and veg suppliers Ted's Veg.

We may once have assumed that plants such as tomatoes and lettuce could only be grown on land, yet now we discover that healthy plant life can grow under the sea. But what if food didn't even need to be grown? What if you could simply 3D-print it? Not only has 3D-printing proved to be viable (there's one available for use in the office space in which we're writing this very book), the materials that can be used to print with have moved on in leaps and bounds from the original high-tech plastics to a mind-boggling array of different materials.

3D-Printing

Once upon a time, the mere mention of the word 'printing' would have made us shudder at the thought of all the hours wasted having to yank bits of jammed paper out of the jaws of a printer, with its angrily blinking lights and beeping error tones.

Today, our reaction would be quite different. We might even

manage a smile, because new technologies mean that our futures
may soon be filled with products, food and even body parts created
using the new-fangled innovation of 3D-printing. It's not actually a
particularly new idea: the earliest prototypes came out in the 1980s,
making them pretty retro (so it's probably only a matter of time before
East London hipsters start wearing them as necklaces). In recent years
developments have really ramped up and 3D-printing now looks set
to revolutionise the medical, manufacturing, motoring, artistic and
catering industries. If you've got enough dosh sloshing around in
your bank account, you could even pick one up and install it in your
home, just for fun.*

Here's how it works:

First you have to create a digital model of the object you want to
print using 3D computer-aided design (CAD) software. Then you
feed this digital template into the 3D printer and leave it to do its
thang: fusing successive layers of liquid-based materials, one on top
of the other, until the object you designed gradually materialises out
of thin air.

Back to the Future II screenwriters may have missed a trick in not predicting the impact of 3D-printing on twenty-first century living, but
the real world hasn't, and here are some of our favourite applications.

Human Body Parts

Bio-printing has been around since the early noughties, when the first
human tissue was created by printing layers of skin cells interwoven
with structural materials. From skin and craniums, to liver cells and
heart valves, 3D-printing is already changing and saving lives.

---

* Your basic 3D printer starts at around £250 these days, although the top-
  notch ones can be as much as £50,000.

In 2014 the first human ears were printed. We have to admit this got us pretty excited, not least because someone actually managed to print off an exact replica of Vincent van Gogh's severed ear using DNA from one of his relatives. But, despite the bragging potential of a van Gogh ear, there are far more serious applications.

One example would be for the thousands of babies born every year with microtia, a condition that means one or both of their ears may be deformed or non-existent. Traditional prosthetic ears can do the trick to fix this, but 3D-printed versions can create far more precise replicas of the real thing, resulting in a much more natural-looking end result. The procedure by which this miracle is achieved is pretty incredible, too. A bio-compatible plastic is printed out in the desired ear shape, to act as a scaffold, and then planted under the skin on the patient's arm or thigh where it remains for one or two months while natural skin grows around it and there you have it – a nice new ear ready to be transplanted to the side of your head.

CHIC FACT: A Shanghai-based company is now offering bereaved families the opportunity to 3D-print any missing body parts of the deceased to make them more presentable when saying their final farewells. They can do an entire face for a few hundred quid.

A Working Gun
This one scared the bejesus out of us, but it's still pretty incredible to think someone could simply click 'print' to create a fully functioning gun. Luckily the home-printing of guns has now been made illegal . . . although one company did share its 3D gun designs online, which predictably enough went viral immediately and ended up being spread all over the internet. Cheers for that guys, greater access to firearms – just what the world needs.

A Castle
It's every princess-obsessed little girl's dream come true: your very own fairy-tale castle. It turns out it's also one big American guy's as

well – Minnesotan Andy Rudenko, who managed to make his fantasy happen by *printing* a concrete castle in his very own backyard. He built a giant 3D printer especially for the job and it took only one month for the castle to be completed. His next big project? Printing a two-storey house. We dread to think how stressful unjamming a concrete printer would be . . .

Stuff in Space

In 2015 when British astronaut Tim Peake first started cruising above Earth in the International Space Station and needed something, he soon realised that it's not quite as easy as popping out to the local shops when you reside in space. Unfortunately, orbital convenience stores are still some way in the future, but luckily for him and other space-dwellers there's now a very simple solution. Once again, it's the trusty 3D printer. Technicians down on Earth can work up the CAD template design for anything Tim might find himself in need of, from hammers to shoes, cameras to guitars and email it over so that the intrepid astronauts can print it from the comfort of their spacecraft.

CHIC FACT: Hikers can capture their most impressive mountaineering achievements in a 3D-printed mini-sculpture of the terrain they conquered and the specific path they took.

Pizza

It's the answer to all our pizza-loving prayers. Someday soon you could be lazing about at home, feeling that familiar pang of hunger and, despite the empty fridge, satisfy the urge to gorge simply by turning on your 3D 'Foodini' printer and whizzing yourself up some tasty treats.

The printer works like a normal 3D printer, except that raw food ingredients are loaded into little stainless steel cartridges, as opposed to the usual concrete or plastics. For the time being these are mostly used in industrial kitchens, but soon pre-packaged capsules should be available to buy for loading into 3D printers to make fresh food at home, such as pizza, burgers and pasta. However, you will still have

to bear the indignity of actually popping it into the oven for cooking, for the time being anyway.

Meat
One of the strangest things to happen in the world of 3D-printing is that companies have worked out a way to print meat – real meat! To pull off this feat they have taken inspiration from bio-printing technology used by the medical profession (as explained earlier). Stem cells can be collected from live animals during biopsies, then replicated *in vitro* over and over until there's enough to be loaded into a bio-printer cartridge. The living cells clump together into living tissue (known as the delicious-sounding 'bionk'), which can then be printed off into any meat form you might care to dine on – steaks, sausages and even burgers.

Creepy as this may seem, the implications could be huge. Not only would countless animals no longer need to suffer the discomfort associated with intensive farming just to keep us fed, but it could also yield great benefits for our health and the environment. The meat industry contributes to the destruction of the rainforest and other habitats in the never-ending search for more fresh pastures, produces huge amounts of greenhouse gasses and has a heavy impact on our collective carbon footprint as it is imported/exported all over the globe. Overconsumption of meat has also copped the blame for contributing to soaring rates of obesity, promoting the development of several cancers and even threatening to wipe out the human race by helping to breed superbugs (see. pp.240–5). 3D-printed meat could solve all of these problems – so it's much more than a flashy fad.

> **CHIC FACT:** A 3D-printed burger currently costs a whopping $300,000.

Other 3D-Printed Products

An acoustic guitar, a flute, antibiotics, camera lenses, a model foetus, a pair of shoes, fabrics, an office building, a lawnmower and a wrench are just a few of the incredible things you can now 3D-print. Plus, as if that wasn't exciting enough, Phoenix-based company Local Motors have now shaken up the automotive industry by creating the world's first 3D-printed car – the Strati. With their plan to open fifty 3D-printed-car micro-factories around the world, helping teenagers to get their hands on their first set of wheels may well have just got a whole lot cheaper.

A world with printed pizza at the press of a button, real-life hoverboards, flying cars, underwater habitats and farms all sounds like the kind of future we were hoping for back in the twentieth century. Yet the ultimate promise of what the future might hold includes the prospect of cheating death.

I Wanna Live Forever

> Will you take what's in my head?
> And erase me when I'm dead?
> *The Offspring* – 'The Future is Now'

Why are so many of us pre-occupied with death? Could it be the uniquely human curse of being born with an awareness of our own mortality, and our difficulty accepting that one day the person we know best in the world (i.e. ourselves) simply won't exist, while life for others still goes on? Or could it be a powerful form of FOMO – the desire to stick around so we don't miss out on seeing what future holds, to look on as exciting new events unfold, to – as artists from Oasis and Queen to the cast of *Fame* have sung about – 'Live Forever'? Many people spend their entire lives dedicated to the quest of beating death at its own game. Some try to achieve immortality by reaching the dizzy heights of fame, others choose to have dozens of children and some, our kind of guys and gals, use science and technology to try and cheat biology.

Cloning, age-reversing beauty products and human-cyborgs (see Chapter 5, 'Paranoid Android') are only a few of the techniques being explored in the multi-million pound quest to sock one to the Grim Reaper. But our favourite of all? Uploading your brain to a computer.

Scientists have been wrestling with various ways to potentially map out, copy and 'upload' our brains for some time. It may one day allow us to live forever, albeit inside a machine and with the ever-so-slightly compromised quality of life that this may entail. After all, our brains are essentially a giant 3D circuit board, just with trillions of connections in this particular biological circuit and considerably more complexity within each connection.

Having a spare copy of your brain lying about could come in handy in a way that goes beyond mere immortality. You could potentially tweak certain areas to ramp up self-confidence or suppress pre-existing phobias, for example. If you got a nasty bang on the head

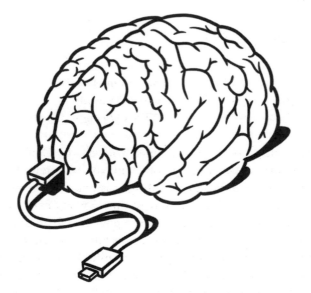

that resulted in some loss of brain function, you could replace the damaged area with an identical back-up – a little bit like syncing your new iPhone with your computer to get all the original info loaded straight back onto it. Come to think of it, 'iMind' has a nice ring to it.

CHIC FACT: There are four known animals that can technically 'live forever'. The Immortal Jellyfish, or *Turritopsis dohrnii*, cheats death by reversing the ageing process and returning to its earlier polyp state before growing up all over again. Flatworms are known for their ability to regenerate, and many scientists believe that as long as lobsters and turtles can manage to avoid disease and the jaws of predators they could live forever, as their internal organs show little sign of degrading over time.

While an actual iMind languishes at the patent office, several bona fide scientists are busy working on genuine solutions. Many believe the first critical step is mapping the human brain. Barack Obama got in on the act in a big way during his time in office, launching the multi-billion dollar BRAIN Initiative in 2013. This has been funding all sorts of inspiring research projects, each taking up a different part of the monumental challenge of capturing the very essence of a human brain.

One company named Brain Backups is trying to use non-invasive brain scanning techniques, such as magnetic resonance imaging (MRI), to map out the entire human brain and all its connections. Eventually, it could allow people to create a basic brain blueprint, storing their own unique map of brain connections and functionality. The trouble at the moment is that even if we could map every single connection between our eighty-six billion neurons and the additional eighty-six billion support cells, we don't have a template in which to store this incredibly complex data set in a meaningful way.

Even if we did get to this stage, the scientists involved reckon an entire brain would require between 1,000 and 10,000 terabytes of storage. That's a mighty expensive hard drive and one you certainly wouldn't want to accidentally leave on the bus.

GEEK CORNER: Trying to draw parallels between the processing power of brains versus computers has come under some perfectly reasonable criticism. While computers are strictly digital, brains are better described as analogue devices. Each neuron has billions of molecular components that can be added, subtracted or moved around to influence the role it plays in a given brain circuit. This is not the case in a computer, making the task of recreating the complexity of brain structure and function incredibly challenging for an inherently digital device such as a computer.

Right now the only way to map a human brain in its entirety is to finely slice it, which of course means destroying it. And did we mention the whopping £1.8 million price tag? Even then the brain map you'd get would only capture its state at a single point in time, which is far from ideal given that brains make new connections and break old ones every minute of every day. However, it is an important first step and one that may prove incredibly useful in several areas of brain health, such as research into the causes and potential cures for illnesses such as Alzheimer's disease and autism.

CHIC FACT: The longest living mammal is the bowhead whale, with one individual known to have survived for a mind-blowing 211 years.

Another team did manage to accurately simulate one second of human brain activity in 2014 using a Japanese supercomputer – like a computer, but much, much more powerful (*and* super!) The 'K computer', the fourth most powerful computer in the world, achieved this feat, but it took forty minutes to crunch all the data once it was collected. The simulation only represented 1 per cent of the total neuronal network of a human brain by replicating 1.73 billion brain wires and 10.4 trillion connections between them. Quite a way to go, then, until an entire brain can be 'uploaded' and replicated, but not necessarily impossible. A far more powerful exascale computer could hypothetically do the full monty. The only problem? Such computers don't yet exist, but computer bods reckon they'll have created one by 2018.

GEEK CORNER: Scientist Aubrey de Grey believes that in twenty-five years' time the technology will exist to allow us to live for a thousand years or more. What's more, he suggests we'll be able to remain looking and feeling as young as we like. The idea is to alter our genomes to include the genetic codes of soil-dwelling micro-organisms that can break down the waste proteins created by our cells that we're unable to fully remove by ourselves.

We may be a long way away from full-on *Matrix* territory, but scientists *have* managed to upload a worm's brain and many believe the ability to upload our own brains is just a matter of more time and effort.

One thing's for sure: our species' fixation on life and death will continue long into the future. Technological and medical advances currently allowing us to live for longer and longer may one day even allow us to live forever. We should be careful what we wish for . . .

Final Thoughts

You've now had a tiny little slice of what life in the future may look like and we're gonna be honest, it looks pretty exciting. Some of it may be focused on clearing up the colossal mess we humans have been making on Earth thanks to a whole heap of earlier nifty innovations gone by. But much of it hopes to make life simpler, easier and a whole load more fun (3D pizza printing, we're looking at *you*!).

As our world continues to become more similar to that we once dreamed of on the silver screen we look forward with anticipation to what our futures may hold. Because if there's one thing that *Back to the Future II*'s predictions of the world in 2015 taught us, it's that we can imagine all we like, and some of it may even come true, but the only way to really know what's around the corner is to be patient and simply wait and see.

2

Psycho Killer, Qu'est-Ce Que C'est?

What is it about killers that we human beings find so intriguing? On the one hand, we're disgusted by them, driven by an overwhelming empathy for the pain and suffering of their victims. On the other, we also find them absolutely fascinating, with the media and the silver screen seemingly besotted by them.

Notorious evil-doers such as Myra Hindley, Ted Bundy and Aileen Wuornos become the subjects of countless films, documentaries and books; a quick Google search of 'Charles Manson merchandise' reveals no shortage of rock 'n' roll T-shirts, hoodies or badges branded with his face. And, more recently, the mass-murdering 'Boston Bomber' was splashed all over the front cover of *Rolling Stone* – an honour usually bestowed upon only the most revered musicians. Yet while the murderers effectively get to see their names up in lights, their victims are usually confined to obscurity. Seems dead wrong, doesn't it?

Is it because these killers do something extreme that we would never dare, or care, to do? Or perhaps it's because they represent something akin – in a much more disturbing way – to an adolescent desire to break the rules, to do the forbidden and screw the system?

Whatever the reasons, our interest in the deadly and the macabre isn't going to go away anytime soon. And it's not just the film-makers,

novelists and newspaper editors who share in this fixation. Scientists are in on it, too.

Over the course of this chapter we'll be looking at some of the weirdest, most shocking and ghoulish scientific research into the darker side of human nature. We'll put zombies, psychopaths, serial killers, cats and even skinny jeans in the dock, but we begin with one of the most talked-about killers of all time.

Jack the Ripper

9 November 1888: the slums of London's impoverished East End. As the foggy Friday morning dawns a horse and cart rolls lazily along a cobbled street. A morning newspaper thumps against a front door, breaking the peace, and there, emblazoned across the front page, is the headline:

Jack the Ripper Claims 5th Victim –
Woman Brutally Hacked To Death

This headline – which appeared in the *London Daily Post* – marked the fifth of eleven savage murders attributed to the person known as everything from 'The Whitechapel Murderer' to 'The Leather Apron', and eventually the globally infamous 'Jack the Ripper'. The victims were mostly prostitutes, the murders always brutal.* Organs were often removed from the victims' bodies, sparking rumours that the killer was perhaps a surgeon or doctor by trade, yet the perpetrator of these hideous crimes was never caught.

---

* It's very likely that he committed at least five more murders on top of the eleven officially attributed to him.

What started out as a series of very gruesome and tragic murders turned into a national police hunt, a Victorian media frenzy, and then nearly 130 years' worth of speculation, false letters, books, films, TV series, murder-site tours, museums and even gory re-enactments at the London Dungeon. Today, Jack the Ripper is a central part of London folklore – he and his crimes have become an industry for some, glorified entertainment for others and an obsession for many. Fuelling society's morbid fascination with this killer is the fact that his identity has always remained a mystery . . . that is, perhaps, until now.

Modern scientific techniques and innovations have begun to shine a light on who London's most notorious killer might have been and, in our eyes, at least, none is more exciting than the efforts of molecular biologist Dr Jari Louhelainen.

Luckily for the podcast-making duo behind this book, Dr Louhelainen agreed to a world exclusive interview on our *Geek Chic Weird Science* podcast so that we could grill him. And grill him we did. Here's what he told us about his part in the Jack the Ripper story:

In 2011 an email came through to Dr Louhelainen suggesting that a long lost piece of evidence linked to an old forensic case had been rediscovered. Dr Louhelainen agreed to meet the owners and, a few days later, there was a knock on the door of his lab. In came a man named Russell Edwards carrying a brown briefcase, who announced with a wide smile: 'I think you might be interested in this.'

Inside the briefcase was, he said, the last remaining piece of physical evidence from the crimes of Jack the Ripper: an old silk shawl. Dr Louhelainen was sceptical at first, thinking: *here's another nutter trying to fake things to claim some glory from London's notorious mystery murderer.*

Regardless, he was convinced to run some tests on the 126-year-old silk shawl – which Russell Edwards claimed had been taken from the scene of the crime when the fourth of Jack the Ripper's alleged victims, Catherine Eddowes, was murdered and gratuitously mutilated in 1888. A letter Edwards brought with the shawl claimed that Amos Simpson, a sergeant on duty at the time, had half-inched* the

---

* For those unfamiliar with such jargon 'half-inched' is Cockney rhyming slang for 'pinched', as in stolen.

bloody shawl from the crime scene and taken it home for his missus, the old romantic that he was: *Oh darling, I love you so much that I stole an old blood-spattered shawl from a murder victim and brought it home for you as a gift.* Lucky Mrs Simpson.

CHIC FACT: Jack the Ripper only ever killed in the early hours of the morning, or at the weekend. Many suspect he worked weekdays and was a bit of a night owl, staying up in the wee hours to commit his crimes. This fits nicely with recent research suggesting that people who stay up late are more likely to be psychos.

This was, of course, well before the days of forensic testing, so Sergeant Simpson may well have thought that there was no harm in taking it, unaware that in years to come it could prove vital in cracking the case. Mrs Simpson, no doubt horrified by its gory stains, never wore it or washed it, and the shawl, along with the story behind it, was passed down through the Simpson family until 1993, when Simpson's great-great-nephew handed it in to Scotland Yard's Black Museum*, where it was eventually sold at auction.

Right from the start Dr Louhelainen asked as little as possible about the shawl's history, to keep him free from possible bias. So, at first he didn't even really know what he was looking for. Soon enough, he uncovered some rather gory details. The shawl was covered in remnants of human internal organs and blood – from the

---

* The Black Museum is a collection of criminal memorabilia kept at London's New Scotland Yard police headquarters.

victim – but there were also some unexpected semen stains, presumably from the assailant himself.

Incredibly, Dr Louhelainen was able to recover DNA from the 126-year-old blood and semen, and set out to match the DNA to descendants of both the victim, Ms Eddowes, and various Ripper suspects. How did he manage to find these elusive descendants? A combination of three years of patient waiting and luck being a lady: a TV show that one of his team happened to catch involved a woman who mentioned that she was a direct descendent of Catherine Eddowes. Amazingly, the mitochondrial DNA (see Geek Corner) in the bloodstain matched that of Ms Eddowes' descendant. Even more incredibly, Louhelainen's team managed to track down a descendant of the sister of one of the Ripper suspects and found that her mitochondrial DNA matched that found in the semen.

And the person that the evidence incriminated was—

Drum roll, please

—Aaron Kosminski.

A Polish barber and a key suspect at the time, Kosminski was sectioned in 1891 and spent the rest of his life in a lunatic asylum. Could the true identity of Jack the Ripper finally have been revealed?

GEEK CORNER: Most of our cells contain huge numbers of mitochondria – energy-releasing, cellular powerhouses which have their own DNA. All mitochondria are inherited from our mothers. This means that mothers, but not fathers, pass their mitochondria on to the next generation. You can trace this particular characteristic strand of DNA along the maternal paths of any family tree. Female relatives will have identical mitochondrial DNA, which is conserved all the way down through the generations. Finding the right chain of female descendants from the victim and the culprit was key to cracking the mystery in this case. However, there is still a small chance that the match is completely coincidental, hence the continuing doubt. Mitochondrial DNA testing has had huge success in other types of detective work, like plotting the trajectories of our ancestors migrating out of Africa.

Even Dr Louhelainen admits this is only the beginning of his work and that he would need to complete many further tests to be 100 per cent convinced. Yet, despite the uncertainty, news that Aaron Kosminski *could* finally have been proved to be the Ripper sent the world's media into a frenzy. Hordes of doubters were soon lining up to criticise Dr Louhelainen's work, far outnumbering those offering praise for a job well done. Some picked apart the story of Amos Simpson, claiming he could never have even been on duty that night. Others poked holes in Dr Louhelainen's methods and accused him of errors – many of which he doesn't deny, although he still maintains they don't affect the rest of the evidence that points overwhelmingly to his conclusion: Aaron Kosminski *was* the Ripper.

All these on-going bones of contention mean that the mystery of who Jack the Ripper is still hasn't been fully cracked and, even today, new theories continue to fly around, with a new 'unmasking' book or article popping up every month or so[*]. One recent theory accused the poet Francis Thompson of being the Ripper. Another writer claimed that his own great-aunt was a victim, murdered along with the others by her husband, Francis Craig. Even *Withnail and I* director Bruce Robinson has got in on the act, pointing the finger at celebrated Victorian music star Michael Maybrick.

CHIC FACT: The moon has long been associated with the capacity to trigger madness. So entrenched was this belief that the word 'lunatic' actually derives from the work 'lunar' i.e. moon-related. In the Victorian age members of the public could pay a penny to see the moon-perturbed nutcases at the lunatic asylum, purely for entertainment value. Sick puppies.

The truth is that the identity of the notorious killer may never be fully known, nor accepted beyond reasonable doubt and new 'Ripperology' theories are not likely to dry up anytime soon. But one

---

[*] A quick web news search of the words 'Jack the Ripper' and 'identity' shows that between July and November 2015 an article proclaiming the real identity of Jack the Ripper appeared every single month.

thing's for sure: if anything can ultimately put all this speculation to bed and answer the question that's perplexed generations of wonderers – *who was Jack the Ripper?* – it will, like so many of life's mysteries, almost certainly come from the wonderful world of science.

While Jack the Ripper was clearly severely mentally unhinged, whether or not he was a bona fide psychopath will probably never be known for sure. But what does being a 'psycho' actually mean? Do all psychopaths dress up like their mothers and stab people in the shower? And is it possible to spot one – even if the one you spot happens to be yourself?

Does Loving Coffee Make You Psycho?

As you read the above title you may well have found yourself peering suspiciously at the people around you sipping on their skinny lattes. But, before you go accusing complete strangers – or your nearest and dearest, for that matter – of being the next Norman Bates, there's more to this than first meets the eye.

The study that gave rise to it involved a questionnaire in which people were asked about their various food and drink preferences. The same people were also asked to do a series of personality tests to discover how they scored on psychometric scales relating to a variety of personality disorders. These included the 'dark triad': Machiavellianism (the desire to manipulate others), narcissism (an overblown feeling of self-importance and craving for attention) and psychopathy (a lack of empathy).

CHIC FACT: Internet trolls have been found to show strong signs of psychopathy, but also sadism, Machiavellianism and narcissism – the combo known as the dark tetrad. The more time spent trolling, the stronger the traits. (*Daily Mail* commenters, we're looking at you!)

It turned out that people who confessed to a penchant for bitter food and drink, including coffee, but also beer, dark chocolate and even broccoli apparently (is broccoli bitter?!), also tended to score

highly on the psychopathy scale and for several other antisocial traits. So what does this actually mean? For now, it's simply that they've noticed a correlation: people who score highly on loving bitter foods also tend to be those who score highly on the psychopathy scale. It doesn't necessarily mean that one causes the other – that indulging in a cappuccino or a cheeky pint will immediately turn you into the next Dexter, or that being a psycho will necessarily make you Starbucks's most loyal customer. However, by that same token, we also can't with absolute certainty that this *isn't* the case. A connection *has* been uncovered and it's up to the next generation of scientists to figure out exactly why and whether or not one causes the other.

GEEK CORNER: Psychopaths may be driven by an irresistible craving for more dopamine – the same neurotransmitter that is released in our brains' reward pathways to trigger the feelings of pleasure that arise when we eat sweets, make high-risk decisions, fall in love or take recreational drugs. A recent brain scanning study indicated that those scoring highly on the antisocial impulsivity scale may produce more dopamine, suggesting that they may feel heightened pleasure when pursuing risky goals, like obtaining money, sex or status under dangerous circumstances.

It also good to remember – before you start lurking around the work coffee machine waiting to accuse its most frequent user of murdering their nan – that the medical term 'psychopathy' isn't quite the same phenomenon as implied by the characters routinely peddled by pop culture. A psychopath is actually someone who is incapable of feeling empathy – they have no intrinsic sense of how other people feel at any given moment. In neuroscience terms they even show decreased activity in the parts of their brains that enable empathy to take place in 'normal' people under circumstances where it usually comes into play*.

They may appear cold-hearted, or lacking in emotion, but that certainly doesn't make them guaranteed killers. In fact, recent studies suggest that lack of empathy combined with other characteristic traits of psychopathy – such as boldness, fearlessness and impulsivity – can in fact make for very successful leaders, be they army generals, tycoons or even presidents (we can think of a fair few . . .).

> **CHIC FACT:** Not all psychopaths are bad. Neuroscientist James Fallon discovered he was a full-blown psychopath while analysing the brains of his own family of dubious morals (with no fewer than seven alleged murderers in the family tree). However, Dr Fallon has never committed a crime, continues to do good work, and is, to all intents and purposes, an upstanding citizen. He does say that he feels the same way about his granddaughter as he does about random people on the street – but, hey, nobody's perfect.

Some studies have found that non-criminal psychopaths, or 'corporate psychopaths' as they're sometimes known, are particularly rife within the business world, especially in higher management roles and positions of greater power. In 2010 one extensive survey found that up-and-coming management executives were four times more likely (at 1 in 25) to be full-blown psychopaths than the rest of

---

* Much of this work has involved driving an MRI scanner into a high-security prison in the United States in order to scan the brains of psychopaths that have been banged up for horrific crimes. You'd have to be one seriously brave neuroscientist to sign up to that research programme!

the population (1 in 100). Then in 2013 a researcher named Kevin Dutton went even further and pulled together a list of professions that certified psychopaths were most attracted to, finding that jobs such as police officers, lawyers and surgeons were high on the list. The number one top job for psychopaths? CEO. Hmmm, hotshot bosses who drink lots of coffee and display cold and psychopathic tendencies? Nah, we've never come across any of those ourselves either . . .

You'd be forgiven for thinking you had enough on your plate trying to spot any coffee-loving, power-crazed psychos in your life, but get ready to be even more paranoid. We've got one more possible give-away for you.

Night Owls More Likely to be Psychos

Ask yourself this – are you a night owl: someone who likes to stay up late and then laze away the morning hours in bed hitting the snooze button? Or a morning lark, who lives by Benjamin Franklin's mantra 'early to bed and early to rise makes a man healthy, wealthy, and wise'?

Anyone falling into the former category had better pay attention, because in 2013 a bunch of Australian scientists found that night owls are far more likely than their morning lark counterparts to display the dark triad traits.

> **CHIC FACT:** Human beings who show strong dark triad traits aren't the only ones more likely to be creatures of the night. Predators in the animal kingdom such as lions and scorpions are also more likely to be nocturnal, so perhaps they share a similar motive.

Now, it could simply be that some of the antisocial activities, and crimes for that matter, that those rating highly on the dark triad are more likely to partake in, can more easily take place in twilight hours. Poor lighting, fewer potential witnesses knocking about and the fact that those who *are* around may well be sleepy and therefore more vulnerable, all contribute to the perfect conditions for petty crimes or even murder. Psychopaths, for example, may simply *choose* to operate at night when they are less likely to be caught.

GEEK CORNER: Psychopaths have great difficulty recognising fear in a person's facial expression. A test of thirty-six children aged seven to ten years old showed that those who scored highly on the psychopathic scale struggled to tell the difference between pictures of scared versus relaxed facial expressions. This might boil down to the structure of their amygdalae: almond-shaped structures deep in the temporal lobes with an important role in spotting danger and triggering fear. They tend to be smaller in psychopaths compared to others.

It's possible that there's an evolutionary link between a love of late nights and the prevalence of dark tetrad traits. People with a tendency towards antisocial behaviours who were morning larks may simply have been more active in daylight hours and therefore got caught out more often, rendering them less likely to pass on their antisocial genes. Meanwhile, those burning the midnight oil may have been more successful in evading detection and so left free to sow their evil oats.

CHIC FACT: It's not all black and white. While there are plenty of examples of evil night owls, such as Adolf Hitler and Josef Stalin, there are also tons of good ones, including Elvis Presley, James Joyce and Barack Obama. Morning larks can be a mixed bag, too: Napoleon and Ernest Hemmingway are two notables who liked to be up at the crack of dawn.

The precise reasons might be unclear, but a link between night-owlish waking hours and psychopathy does nonetheless exist. Late-night lovers, don't despair. Before you run out and book yourself an appointment with a psychiatrist, it's not all bad news. Night owls are also said to be more likely to be more intelligent, extroverted and creative – artists, poets and musicians, for example, are often fans of the long, dark night. On the other hand, morning larks are more likely to be accountants or civil servants. Psychopath or accountant? Neither seems all that appealing!

Psychopathy does not a murderer make. Yet lacking the capacity to feel another person's pain surely makes murder considerably easier to stomach. We certainly wouldn't want any psychopaths to get their hands on the next story because we may have uncovered the perfect murder.

The Perfect Murder

> Countess Vera Rossakoff: And what of the perfect crime?
> Hercule Poirot: It is an illusion, Countess.
> Agatha Christie's Poirot: 'The Double Clue' (1991)

Is there such a thing as the perfect murder? It's a question that has occupied the minds of brilliant crime writers, film-makers, detectives and criminals alike for many years. Now it seems a method may have been uncovered that has the potential to outwit even the great Hercule Poirot (*mon Dieu!*).

It's a ploy worthy of our finest screenwriters and one that many readers may find familiar, especially fans of the TV series *Homeland*. In series two [spoiler alert!], in an episode aptly titled 'Broken Hearts', Vice President Walden is murdered by a terrorist who was nowhere near him at the time. The villain was able to remotely hack into the Vice President's pacemaker from some far away location and switch it off. The best (or worst, if you're in Walden's fan club) part is that, to everyone else, it just looked like the poor sod had suffered a heart attack. No one suspected foul play, nobody was charged and the murderer got away scot-free. We watched wide-eyed and then turned our TVs off, safe in the knowledge that this was only the stuff of fiction, right?

The plot itself was actually inspired by an article that one of the screenwriters chanced upon. It shared news of real-life research that had uncovered a potentially fatal flaw in modern pacemakers and a variety of other life-sustaining medical devices. Computing security researchers argued that, as pacemakers are often wirelessly controlled, it could leave people vulnerable to potentially fatal meddling from hackers. You'd only need to guess the password to hack in, which the researchers discovered was, more often than not, left on the factory default setting.

'Why on earth would pacemakers be created with a deadly flaw as easy as remote access?' you may ask. Most modern pacemakers implanted into patients are designed to connect to the internet. This enables them to be monitored and adjusted remotely by cardiologists without needing to crack open your chest every time a tweak is required. That all makes sense when your kindly doc is keeping a careful eye on your heart and making sure it's beating nice and steady. Not quite so comforting when you remember that for every gazillion altruistic doctors out there may be another Harold Shipman.

Some of us struggle to fully give our hearts to the ones we love the most, so giving your doctor full access to and control of your heart clearly requires a great deal of trust. That's before you even start to consider the many tech-savvy total strangers out there champing at the bit to try out their hacking skills on a new challenge. And while Hollywood stars like Jennifer Lawrence may face embarrassment when their personal data gets hacked into, that's nothing compared to the potential risk to the hundreds of thousands of people across the globe currently wearing pacemakers.

Luckily, tampering of this deadly nature could prove harder to achieve than it first seems. While around one million people a year are fitted with pacemakers, thankfully most of them wouldn't die even if their pacemakers were turned off, because they are often fitted for non-life-threatening heart conditions. However, the threat was still enough to motivate the former Vice President Dick Cheney to disable the wireless function of *his* pacemaker after advice from his doctor.

Other remotely accessible medical gear is also at risk, with more than twenty devices currently being investigated by the *real* Homeland Security. These include insulin pumps, defibrillators, anaesthetic pumps and all sorts of other implanted devices that can be connected to wirelessly and controlled remotely.

> **CHIC FACT:** Medical devices are not the only pieces of technology that can be hacked with potentially fatal consequences – cars can be hacked, too. In July 2015 one car company had to recall 1.4 million cars after hackers demonstrated that its vehicles could be remotely hacked and forced off the road by someone tinkering with them via their wirelessly connected entertainment systems. Scary!

Security experts are starting to clock on and clamp down on these flaws. New tracing methods have also been developed to help pathologists detect signs of foul play. But until it's completely watertight security-wise, nobody knows if or when the first murder by remotely accessed pacemaker will happen. That is, of course, if it hasn't happened already, because if it had, none of us would be any the wiser. And that, friends, is not just what makes it terrifying, it's what makes it *the perfect murder.*

We now know what to watch out for when it comes to spotting a psycho. We'll be keeping a beady eye on those chocolate-, beer- and coffee-loving, high-status professionals with late-night tendencies and computer hacking skills. But what about the other less obvious threats out there? Like . . . kittens?!

Are Cats Messing with Our Minds?

Cats get a hard time, don't they? They're bad luck if they cross your path. Bad luck if they're black. Bad luck if you owned one in

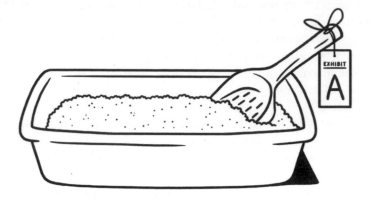

sixteenth-century Britain and didn't fancy being burned at the stake as a witch.

Now, to top it all off, modern-day scientists are accusing them of potentially contributing to mental illness, suggesting that owning a cat in early childhood could have links to developing schizophrenia in later life.

The research in question looked back at a detailed survey taken by 2,125 families in 1982 as part of a scheme run by the National Institute of Mental Health. They found that 50.6 per cent of those who went on to develop schizophrenia had owned a furry feline in childhood. Two other similar studies that took place in the twentieth century showed almost identical results.

This may of course just be a coincidence. As we never tire of reminding ourselves (and you), identifying a correlation between two things doesn't necessarily mean that one actually caused the other, only that it *might* be the case and should probably be investigated further. One particular theory has gained a lot of traction over the years. It proposes that a parasite, *Toxoplasma gondii*, could be to blame. *Toxoplasma gondii* (or simply *T. gondii*) is a nasty little parasite found living inside many cats. It can be passed onto humans via kitty litter and causes nasty microscopic cysts when it gets into our brains. These can become activated later in teenaged brains, affecting neurotransmitter release and potentially, it seems, triggering schizophrenia.

T. gondii has also been linked to miscarriages, blindness and even death, but as the parasite only passes to humans via cat poo, the general advice is simply to avoid kitty litter at all costs if you're immunocompromised or pregnant. Healthy people with strong immune systems can generally keep the parasite in check, but for the immunocompromised, like those with HIV, it could mean big trouble – anyone else remember what happened to poor old Tommy in *Trainspotting*?

GEEK CORNER: Cats aren't the only animals responsible for getting *T. gondii* into the human brain. The parasite is usually passed to cats via their old adversaries, rats. Once *T. gondii* is in a rat's brain it renders it unafraid of cats by removing the rat's natural fear of the smell of cat's urine, even making it feel an irresistible attraction towards it instead. The cunning parasite makes the rat much more likely to be eaten by a cat and thereby gain access to its digestive system. This is precisely what *T. gondii* needs, because inside a cat's gut is its favourite place to reproduce and flourish, having its offspring dispersed all over the place via the cat's poo.

But before you panic and start planning to get rid of poor old Tiddles, just remember that more than 30 per cent of the people in the Western world own cats, and only 1.1 per cent of them have been diagnosed with schizophrenia. With the strong genetic component in this particular mental illness it's unlikely that simply owning a cat could cause schizophrenia. In fact, it may well be the other way around – children with a predisposition to mental illness may simply have a stronger desire to own pets; after all, owning a pet has been shown to help reduce stress levels, ease loneliness and even reduce risk of heart attacks.

CHIC FACT: Cats have been responsible for many memorable inventions. Most people know the story of 'catseyes', whereby a cat's glowing eyes at night inspired Percy Slaw to create the reflective roadside lights so ubiquitous today, but did you know that getting a static shock from his cat inspired Nikola Tesla to start to investigate electricity?!

Does that mean our furry feline friends are once again being scapecatted? Plenty more research needs to be done to know for sure.

Speaking of messing with our minds, have you ever wondered if there really are zombies walking among us? You're not the only ones . . .

Are Zombies Real?

Did you think that zombies were safely confined to films and nightmares? Think again. Some cultures, in Haiti, for example, are convinced that the walking dead are the stuff of real life.

For many years, macabre tales of people being brought back from the grave and kept in a trance-like state were rife among vodun cultures (i.e. those who believe in voodoo). But how reliable are they? And has science ever put these stories to the test?

One of the most famous examples of a 'zombie' was the mysterious case of Clairvius Narcisse. In 1980 Clairvius walked into his village and greeted his sister. So far, so pleasant. However, his sister Angelina started to scream, very, very loudly. It turns out her brother had died eighteen years earlier, in 1962, and had been buried in a nearby cemetery. His death was even confirmed by not one, but two, American doctors. How in the name of the heebie-jeebies was he walking and talking right in front of her that very day?

Clairvius, whose identity was confirmed by a series of rigorous tests, claimed that he remembered being buried and was conscious yet paralysed throughout the whole ordeal. Creeped-out yet? Brace yourself. It's about to get worse. He even said he remembered the terror of the nail going into the coffin and hitting his forehead with a scar on his head to (sort of) prove it. That night, he said, he was brought back to life by a *bokor* – a voodoo witch doctor – who kept

him drugged up and enslaved on a nearby sugar plantation, forcing him to work there alongside other mindless zombies for eighteen years. Hands up who's up for cremation after reading that?

CHIC FACT: Many Haitians believe that if you feed salt to a zombie you will free them – they won't become a living person again but their body will return to rest in its grave.

Fact or fiction? You decide. His case, however, was not one of a kind – plenty of others came forward with similar stories – so one curious scientist, Wade Davis, decided to travel to Haiti to investigate for himself. He was introduced to Marcel Pierre, a voodoo witch doctor seasoned in the ways of black magic. The scientist managed to convince him to reveal how to prepare 'zombie powder' – which locals believed could cause people to appear dead, and later be turned into stupefied slaves.

Davis ran tests on several vials of the strange potion and found that each one varied slightly, but most consisted of different blends of bits of toads, sea worms, lizards, tarantulas and, most chillingly of all, crushed-up babies' bones. The ingredient that really caught Davis's attention, however, was dried pufferfish. That's because it contains an incredibly powerful and toxic nerve poison called tetrodotoxin that lovers of Japan's delicacy *fugu* will be familiar with. Prepared properly by skilled, licensed chefs and *fugu* is delicious; sliced wrongly and it is absolutely deadly, as one unfortunate victim in an episode of *Columbo* found out the hard way.

CHIC FACT: Tetrodotoxin is so-named because pufferfish, from which the toxin derives, have four teeth (*tetras* – four; *odontos* – teeth) fused together into a kind of beak for crunching coral and the like.

Could the nerve impulse-blocking tetrodotoxin be the key to turning ordinary people into zombies? After all, it *is* a neurotoxin thousands of times more poisonous that cyanide. Davis's team ran more tests, rubbing the 'zombie powder' into the shaved backs of some poor unsuspecting rats, which promptly slipped into comas. To all intents and purposes the rats seemed dead, but scans nonetheless found faint brainwaves and weak heartbeats. Bingo!

CHIC FACT: Tetrodotoxin is not strictly limited to the pufferfish. A whole host of other marine creatures also produce and store it in their bodies, presumably to discourage predators. These include parrotfish, triggerfish, goby, cod (eek!), sunfish, starfish, the horseshoe crab, flatworms, ribbon worms, arrow worms, sea squirts, Japanese cone shell and even a few cute little frogs.

It was enough to convince Davis that, like master anaesthetists, voodoo doctors could carefully measure out just the right dose to make a person appear dead to everyone around them and then could later recover them from their graves. He believed *bokors* (sorcerers) would dig them up in their zombie-like state and then enslave them by giving them regular doses of a poisonous plant called *Datura stramonium*, to keep them in a mindless state of delirium and confusion. *Day of the Dead* – eat your zombie-loving hearts out!

So had science really proved that black magic really works? Davis's theory was picked apart by scientists for years as they struggled to recreate the results of his tests. Many pooh-poohed the whole theory as a load of old black magic cobblers*. But Davis did write an incredible book about his experiences called *The Serpent and the Rainbow*, which went on to become a huge blockbuster film directed by Wes Craven. Many credit this movie for bringing zombies to the forefront of the horror genre, spawning the legions of zombie films that followed.

The horror stories of people being killed, brought back to life and kept in a trance-like-state to keep them enslaved as zombies persist today, in places as varied as Haiti, South Africa and the French West Indies. And while sceptics and fans of Davis's work continue to lock

---

* 'Cobblers' is Cockney rhyming slang for 'balls' – as in nonsense: cobbler's awls . . . balls!

horns, rumours that he eventually sold the concoction to Michael Jackson who then forced his dancers to rub the zombie powder into their skin before performing 'Thriller' are, as yet, completely unconfirmed.

GEEK CORNER: Tetrodotoxin (TTX) blocks the sodium channels that enable electrical messages to spread along the wire-like axon of brain cells. The reason it is such a dangerous toxin is that it blocks a fundamental component of every single neuron in the body and brain. If you eat a pufferfish that hasn't had all its TTX removed adequately then it's game over. First your lips and mouth start to feel numb, which soon spreads to your face and extremities. This is followed by progressive paralysis of other body areas, including vital organs such as the heart and lungs, with death following six to eight hours later. Nasty. That said, for the neurobiologist it is an invaluable tool and also has great medical potential as a therapeutic drug for the induction of anaesthesia and possibly pain relief, too.

For human beings there's little more terrifying than the idea of zombies: creatures brought back from beyond the grave to attack us, eat our flesh and turn us into mindless creatures just like them. But what about the rest of the animal kingdom – are there real-life zombies lurking out there among the birds and the bees?

Zombie Kingdom

Say the word 'zombie' and Frankenstein's Monster, Simon Pegg's *Shaun of the Dead* or even The Cranberries (remember *that* song) might well pop into your head. But after reading this you might start thinking of creepy-crawlies as well, because in scientific terms zombies are already all around us; they just happen to take on a very different appearance from what we might have imagined.

Zombie expert (a job title of which we are a tad envious) Frank Swain agreed to be interviewed by yours truly for a Halloween special of our *Geek Chic Weird Science* podcast. He mentioned that while film buffs may see zombies as flesh-eating, walking-dead humans, for scientists the reality is a whole different story. In many scientific papers the term 'zombie' is used – albeit slightly tongue-in-cheek – for any organism whose behaviour is modified or controlled by another creature.

How many brain-jacking organisms are there? Quite a few, as it turns out! Parasites of various forms find devious ways to get into the body of other organisms and adapt their new home into an environment that better suits their needs. A bit like an uninvited guest waltzing into your living room, rearranging the furniture, refusing to leave and then taking over your mind. Sort of.

The poor unsuspecting creatures that fall victim to these zombie-makers can be critters, or, even most shockingly, humans – cue dramatic 'duh duh DUH'.

Here are a few of our and Frank Swain's favourite examples:

The Emerald Cockroach Wasp

This dastardly little beasty takes control of cockroaches by driving its stingers deep inside their brains. The roach then becomes very placid and easily coerced, showing no urge to run away, nor do very much of anything for that matter.

Here's the clever/mean bit: this zombification allows the wasp to drag the cockroach somewhere quiet, lay eggs on it and then cover it in little stones. The roach seems quite happy in its trance-like, tranquil state, but meanwhile all sorts of grim stuff starts happening. The eggs hatch and the larvae burrow into the cockroach and start to eat it from the inside out. Yum! Eventually the baby wasps burst out of the cockroach *Alien*-stylee as fat, happy, fully-fed adults. The cockroach, of course, has been eaten alive and is no more. What a way to thank your gracious host for a generous meal.

GEEK CORNER: Parasites don't even need to get inside the brain of their host to induce the changes in behaviour they desire, as long as they can infect neurons in other body parts. This can trigger an immune response in the central nervous system, which in turn can change neuronal and hormonal signalling to impact on the creature's behaviour in a way that benefits the parasite.

Zombie Fungus

Poor old ants. Not only do they have to contend with the insect-crushing fingers of toddlers and the magnifying glass-focused sunlight of sadistic little boys – not to mention all sorts of more conventional predators – now they also have to watch out for being zombified by a fungus.

If a spore of the tropical fungus *Ophiocordyceps unilateralis* touches an ant things go tits up for the little critter very, very fast. The spore germinates, penetrating right inside the ant's body through tiny holes in its exoskeleton. There it makes itself at home, growing and extending its tendrils and taking over plenty of the ant's soft tissue. The most gruesome part is that the fungus takes over the ant's flesh, but leaves the vital organs intact, effectively keeping it alive for as long as possible while gradually destroying its insides. And there we were thinking Hannibal Lecter was a nasty piece of work.

When the time comes for the fungus to sprout and produce its own spores, its long thread-like branches finally pierce the ant's brain, poisoning it and ultimately taking it over completely. By this point the ant is its own walking coffin and, through no will of its own, will find itself leaving the colony, climbing a plant, and then – freakily always around noon – clamping itself around a leaf, before *pop!* brand new fungal spores spring forth from its head, floating off to zombify some other unsuspecting ant.

CHIC FACT: A list of the safest countries to be in during a zombie outbreak (based on such criteria as population, police presence and topography) has been published. Australia is top of the list, followed by Canada, USA, Russia and Kazakhstan.

The Hairworm

You thought Jiminy Cricket was safe from the perils of zombification? Not while the hairworm exists.

This teeny, tiny parasitic worm makes its way inside the grasshopper and pumps it full of a cocktail of chemicals. The proteins it releases sabotage the grasshopper's central nervous system, effectively brainwashing it and compelling it to commit suicide by leaping into the nearest puddle, stream or pond. Poor old Jiminy drowns, while the hairworm, which needs the water to breed in, thrives, munching down on its host until its offspring are ready to emerge.

These are just some of the ghoulish and ghastly examples of insects being brain-jacked, but here comes the crunch: what about us humans?

Human Beings

According to Frank Swain there is always the potential for us to be controlled. Have you ever had worms? It's OK, you're among friends. If you have, you'll know full well that one of the symptoms of worms is an itchy bum. You don't want to scratch it, but you feel compelled to, and in so doing you touch the worms and their eggs, helping to spread them around to other potential juicy hosts. It's a classic example of parasites making us do something that helps them thrive, with us having no real say in the matter. Several other micro-organisms also affect our minds and can even impact our mental health – remember *Toxoplasma gondii* and its possible link to schizophrenia?

There are plenty of creatures and people among us (maybe even us!) walking around blissfully unaware of a parasite deep inside that may, in some way, shape or form, take control of their thoughts or actions. They may not be in the guise of the archetypal horror film creature we know and fear but, in the minds of many scientists, zombies are certainly among us.

CHIC FACT: A 'secret file' called 'CONOP 8888' was uncovered in the Pentagon's computer network, revealing a zombie survival plan in case of an outbreak. The detailed file – which the Pentagon later claimed to be no more than a light-hearted training exercise – lists the possible zombie threats including: radiation zombies, pathogenic zombies, evil magic zombies and even chicken zombies (which is exactly what it says on the tin: chickens that have become zombified). The plan includes orders to 'kill all non-human life on sight' and to 'target all the main body and holdout vectors of the zombie influence contagion'.

Any good horror book or film would give ample warning to be afraid of zombies, but hands up anyone who's scared of skinny jeans? Turns out you may have something terrifying loitering in your wardrobe . . .

Could Your Skinny Jeans Kill You?

It may sound like every hipster's worse nightmare but in 2015 medics genuinely reported that skinny jeans *could* pose a threat to life and limb. And, no – we aren't envisaging some kind of Revenge of the Killer Denim Brigade with armies of crazed zombie jeans stalking the streets and murdering innocents. We're talking about a case study in a proper medical journal revealing that wearing skinny jeans could severely damage your health and, in the worst case scenario, be fatal.

It all came to light when a thirty-five-year-old Australian woman collapsed after wearing her skinny jeans while helping her friend move house (who was it that said no good deed goes unpunished?). She'd been squatting for prolonged periods of time, presumably while helping her friend pack boxes and lugging them to the removal truck. This, combined with the tightness of her trousers, meant that she badly damaged the muscle tissue in her lower legs as well as the nerve fibres that send electrical messages to and from muscles below the knee. She had effectively cut off the blood supply to her calf muscles and caused such severe swelling that two of the major nerves that controlled the muscles of her lower leg became compressed and dysfunctional.

CHIC FACT: Skinny jeans have been accused of causing all manner of ills, including twisted testicles, heartburn, infertility and making you look like a poser.

On her way home later that evening her feet, and in particular her ankles, simply stopped working properly and she fell over, temporarily paralysed. It took several hours for her to drag herself to the side of a road to flag down a car and she ended up being hospitalised for four days. Medical staff said that if she hadn't had such swift treatment she could have ended up with 'permanent nerve damage [affecting her ability to walk] and that the muscle damage could have led to kidney damage'.

CHIC FACT: Blue jeans were banned from some schools, restaurants and theatres in the 1950s as they were seen as a sign of rebellion. James Dean almost certainly helped to promote that image when he wore blue jeans in *Rebel Without a Cause*. But these days anthropologist Danny Miller reckons more than half the world's population wears jeans, in almost *all* known cultures.

Wondering how muscle damage could lead to kidney problems? Well, when muscle tissue starts to die off it releases huge amounts of protein that can damage the kidneys; part of the reason the woman was in hospital for so long was to keep her on an intravenous drip to flush the excess proteins out of her system. Had she collapsed in a more remote area, further from the nearest road, she might not have had the strength to drag herself all the way into the path of a rescuer and could have died of exposure.

There we have it: a genuine case of a woman's tight-and-skinnies having the potential to cause her untimely demise, but what about the rest of us skinny-wearers? Do half the bands in Camden (and one of your authors) need to throw out their wardrobes? Not quite. Our advice is simply this: use your head. If you're going to be doing some serious squatting, heavy lifting, or both, you should seriously consider opting for looser-fitting legwear. Save the skinnies for when you're doing your best impression of The Ramones.

It turns out that the world really can be a scary place, with all those psychopaths, skinny jeans, zombies and psychosis-inducing kittens on the loose. But could fear itself actually kill you?

Can You Die of Fright?

We've all said it more than once, haven't we? Maybe on a roller coaster, after watching a horror film, or when a friend jumps out at you from a dark corner:

I was scared to death.

But is there any truth behind this old adage? Could we ever be so frightened that we actually die?

The answer in a nutshell is: yes. Although – quick caveat here before this becomes some sort of weird self-fulfilling prophecy and you are scared to death by that fact – it really does depend on how fit and healthy you and your heart are; or aren't, as the case may be.

It's all to do with something known as the 'fight or flight response': nature's way of protecting us by unleashing almighty doses of adrenaline and noradrenaline in response to a potential threat. The surge of these substances into our nerves and bloodstreams is an automatic reaction to something we perceive as potentially harmful. It causes our breath to quicken, our hearts to beat faster, our livers to release sugars from stores, our skin to go pale and we feel a jolt in our stomachs. Why? Because all that lovely glucose- and oxygen-enriched blood is diverted away from organs that don't help us in an emergency

(like our bellies and skin) and into our muscles and brain, which do – so we can either sprint away or stay put and get those fisticuffs going.

This rush of hormones makes us stronger, faster and more quick-witted than we would otherwise be, which can result in some seriously mind-boggling feats of heroics. Ever seen a video of a feeble-looking person lifting a car off a trapped child?*

It can also lead to some not-so savoury results, such as road rage, high stress or even heart damage. If the heart in question is not in very good nick in the first place, then the results can even be fatal. For example, the surge in adrenaline could disturb your heart's rhythm, causing it to pound so fast that it gives up the ghost completely or triggers a deadly arrhythmia whereby the heart muscle contractions get all out of sync and can't pump blood properly†.

GEEK CORNER: The chemical structure of noradrenaline is pretty much the same as adrenaline. A methyl group (a nitrogen atom with three hydrogens attached) at one specific location is simply swapped for a single hydrogen atom. Both are released from the adrenal glands during a fight-or-flight scenario, but noradrenaline is better known for its role as the primary neurotransmitter of the sympathetic nervous system. It bridges the gap between the end of the neuron and whichever organ, gland or muscle that particular brain wire is connected to. When the brain ramps up activity in the sympathetic nervous system, electrical messages are rapidly pinged off to the eyes, heart, lungs, stomach, liver, intestines, kidney, bladder, muscles and skin, to trigger the release of noradrenaline, which switches the organs into a state appropriate for dealing with the danger.

---

* Under such extreme circumstances, the usual reflexes that kick in to prevent our muscles from producing so much force that we do damage to muscles and tendons are switched off. (see p.274–7)

† An arrhythmia is often the root cause for those athletes who, completely unaware that they have a slight thickening in a heart chamber, suddenly drop dead on the sports field mid-sprint.

You may wonder why, if fear itself can be so deadly, we evolved to feel it at all? Well, most of the things we're scared of today posed the biggest threat to our ancestors back in the day: wild animals, deadly insects or dangerous environments. People afraid of these types of threats tended to avoid them and were more likely survive, reproduce and pass on their fearful genes – kudos to Darwin for that insight. But, hang on a minute: isn't this supposed to be a gory, gruesome, psycho- and zombie-ridden chapter? What about more traditional, contemporary sources of fright? Could someone jumping out of a cupboard in a dark room dressed as a ghost actually kill you as a result of the aforementioned heart issues? The answer is: unlikely, but it can and does happen.

After all of our worrying about monsters and men, in words famously uttered by Franklin D. Roosevelt, perhaps the only thing we really have to fear is fear itself.

Final Thoughts

Phew. Scary world isn't it? But perhaps also quite an exciting one. Without zombies, coffee, kittens and – dare we say it – skinny jeans the world could be a pretty boring place. For a life devoid of thrills, chills and perils we may as well wrap ourselves in cotton wool, lock ourselves in a little room and slowly wait for the inevitable to happen. Life's a party, and all good parties come with a few risks, isn't that half the fun?

Our advice? Don't worry too much, as we've seen too much stress can actually put you in more danger than the thing you're afraid of. And be grateful that, for the most part, the really macabre hazards in life, such as monsters and psychopaths, are usually confined to the cinema or, more often than not, simply to our own imaginations.

3

I'm Gonna Send Him to Outer Space, to Find Another Race

The possibility of discovering intelligent life outside of our planet has captured the minds, hearts and imaginations of scientists, astronomers, sci-fi writers, artists and thinkers since the time of ancient Rome. As far back as the first century BC Lucretius wrote in his poem 'On the Nature of Things':

> Nothing in the universe is unique and alone and therefore in other regions there must be other earths inhabited by different tribes of men and breeds of beasts.

Shift forward a couple thousand years and our growing knowledge of astronomy casts ever more light on the extraordinary vastness of the universe. What began as the speculations of a few eccentric scientists, way back when togas were all the rage, has now become a wholesale media obsession with the possibility of alien life. Today, the combination of some incredible innovations, new technologies and brilliant minds means that we have more answers than ever before, and also, of course, more questions: How could our own world be affected by the incredible advances in space discovery? Will we ever live on the moon, or even Mars? And, most importantly of all . . . will we ever find any aliens?

The Hunt for E.T.

We'd be willing to bet that if we gave you a map of your own country you'd quickly be able to point to wherever you call home. Likewise, if we handed you a nice old-fashioned globe you'd probably easily be able to pop your finger on your country of origin. And most, but certainly not all, of you would be quite comfortable describing where our planet is in the solar system. Third Rock from the Sun is quite a memorable address to work from, after all.

It's once you start asking where our planet is in our galaxy that large numbers of people start to go cross-eyed. The number of people eagerly raising their hands to offer an answer takes a further nosedive when asked if anyone knows where our galaxy fits in with all the others out there in the universe. If this last task seems like child's play to you then, while the rest of us thrash about like koi carp in a bucket, you can give yourself a pat on the back for being a clever bean. You can probably count yourself among a handful of the most elite astronomers, planetary scientists and cosmologists in the world; Stephen Hawking is a good example of someone who'd know only too well.

> **CHIC FACT:** Many people claim to have seen, been abducted by and even hooked up with aliens. One of the most bizarre reports was by Brazilian farmer Antonio Villas Boas who said barking aliens took him away, covered him in gel and then promptly mated with him. In 2003 a Harvard study found ten other people who insisted under hypnosis that they had been abducted by aliens for sexual experiments.

In 2015 the genius we all love to love, the aforementioned Professor Hawking, launched Breakthrough Listen, the biggest ever search for intelligent extraterrestrial life with the words:

... in an infinite universe there must be other occurrences of life.

If Prof. Hawking says it's so then we'd be willing to bet our last Rolo on it, and right now the chances are better than ever that we find some alien life sooner rather than later. Is that because Mulder and Scully are finally back and determined to root out E.T.?

Sadly not. The real reason is that we used to think there were around a billion Earth-like planets in the universe – so that's a billion lumps of rock that may have suitable conditions for harbouring life that we'd need to explore, a slightly tricky task considering how very far away so many of them are. However, new research suggests that there are actually billions of Earth-like planets much closer to home, in *our very own galaxy*, the Milky Way! Next time you get the chance to gaze up into the sky on a clear night, well away from light pollution, have a really good look at the giant smear of light that illuminates the sky with all its celestial splendour – that's the Milky Way.* With the newly found billions of Earth-sized planets in that relatively narrow patch of space, that's an awful lot of bites at the cherry of creation. Right on our galactial doorstep.

CHIC FACT: The earliest known UFO sighting was in 1450 BC, when a group of Egyptians claimed to see bright circles of light in the sky.

The SETI (Search for Extraterrestrial Intelligence) project now feels considerably less like hunting for a needle in a haystack and more like hunting for one in a single bale. Good thing, too, because they've been searching the haystack for half a century already with a grand total of ZERO aliens spotted, sorry to disappoint any of you hoping for a Roswell-style revelation here. And just to make it clear – SETI is a quest to find *intelligent* life, meaning any life form that hasn't evolved enough to be able to send and receive its own telecommunications can jog on. Fussy, aren't we?

Back to Breakthrough Listen, which will search for E.T. in our

---

* It can be hard to imagine, but the Milky Way is an unimaginably large number of stars packed into a flat disc shape, like a vinyl record of astronomical proportions. When we look at the Milky Way from Earth, we are looking from near the edge of the record towards its middle, and where this bright strip of stars meets the night sky, on either side we're seeing the upper and lower surfaces of our spinning galaxy.

new and improved hay-bale search-parameter. The project's first job is to point some seriously powerful radio and optical telescopes at the most promising spots in our cosmos, one after the other, for over a thousand hours per year. It will scan the centre of our own galaxy, the hundred closest galaxies and the million closest stars over the next ten years, and in doing so this SETI project will collect more data in one day than in a whole year of any previous search*. It's essentially a little like the SETI researchers have been told they can use the internet from home rather than hunting through endless books in a library in order to write an essay.

CHIC FACTS: If YOU want to join the hunt for E.T. you can download software from the SETI@home project (setiathome.ssl.berkeley.edu), which monitors radio signals from space to help search for alien signals, onto your laptops to get involved from the comfort of your own home. Hundreds of thousands of people already have.

The price tag for this monumental project is a mere $100 million. A huge wad for many, but a small price to pay for the answer to what Hawking described as the biggest question we human beings face.

A word of caution, though: we should be careful what we wish

---

* And no wonder – it will cover ten times more of the sky with fifty times greater sensitivity, amounting to the deepest, widest search for electromagnetic communications than any previous effort. It will also scan five times more of the radio-wave spectrum, one hundred times faster, than ever before.

for, because if we do find intelligent life we'd better hope that we get lucky and find a peaceful mob*. There really are no guarantees that any intelligent extraterrestrials we stumble across won't hold us humans in the same low regard as many of us do a small insect – crushing us underfoot without a smidgeon of guilt. Or they could eat us. Or torture us. Or peel off all our skin.

Come to think of it, we may actually be OK with being totally ignorant of any intelligent beings in our galaxy. Not wishing to be labelled wimps or anything, but given all the gory alien-encounter outcomes dreamed up by many of mankind's greatest works of science fiction, we reckon being a large fish in a small pond may not be so bad after all.

> **GEEK CORNER:** Millions of people say that they've experienced alien encounters. Yet psychological tests on those who claim to have been abducted by aliens rarely reveal mental illness or psychosis. One common explanation is that many people (over 60 per cent of the population, in fact) suffer from sleep paralysis, a terrifying experience whereby the body is paralysed (as is normal in the REM cycle of sleep) but the mind is semi-, or fully lucid and aware of its surroundings. For a few of the sufferers this paralysis can also be accompanied by hallucinations – auditory and visual – which can include seeing bright lights, the feeling that someone is holding you down and choking you; not to mention an eerie sense of being watched.

With the search for intelligent life in full swing our next question is: should we ever find E.T., what would he or she look like?

---

* The Breakthrough Listen project is simply a search, it won't actually send any messages or communication to anything 'out there', an important distinction for scientists nervous about making contact with potentially dangerous and powerful other lifeforms. As Hawking said, 'a civilisation reading one of our messages could be billions of years ahead of us ... and may not see us as any more valuable than bacteria'.

What Would Aliens Look Like?

It's one of the most iconic scenes in movie history. A little boy named Elliott is fast asleep in a chair in his back garden. He's all wrapped up, snug as a bug in a rug, a torch in his hand, still switched on and shining a light into the misty night sky. A crash of metal awakens him and his eyes blink open to reveal a small, tubby creature – with long arms, large feet, big belly and cuboid head cautiously emerging from the garden shed.

Speechless and wide-eyed, Elliott whispers, 'Mom! Mom! Michael!' as the little alien waddles towards him, slowly stretching out those famously long, bony fingers, to drop a peace offering into Elliott's lap – a handful of chocolate M&Ms.

This little E.T. became the epitome of alien life forms for many years to come, but other writers and film-makers, past and present, imagined extraterrestrials as anything from little green men (à la Rudyard Kipling's *Puck of Pook's Hill*, way back in 1906) to human shape-shifters (à la *Men in Black* in 1997). But what would aliens, should they exist, actually look like?

> **CHIC FACT:** E.T.'s face was created by combining the looks of the American poet Carl Sandburg, Albert Einstein and … a pug! This may be related to the fact that Steven Spielberg originally dreamed up the E.T. character as an imaginary friend to help him cope after his parents' divorce in 1960.

One author and evolutionary biologist reckons he has the answer, and it turns out he thinks they'd look pretty similar to *us*. Professor Simon Conway Morris argues that it would be thanks to something called 'convergent evolution', a theory suggesting that, as they evolve, different species start to independently develop similar traits to each other over time in order to survive in similar environments. One example of this is flight, with bats, birds and certain insects all developing this enviable ability, despite not having a common flying ancestor. Another is opposable thumbs, which monkeys, lemurs and humans have all developed.

Professor Morris argues that in order to have a good chance of creating life, a planet must share certain conditions with Earth. At least *some* of the Earth-like planets that have been found out there, such as the recently discovered Kepler-186f* should support some sort of extraterrestrial life that has evolved in similar conditions as us and therefore will be a bit like us.

GEEK CORNER: Why are we so convinced that there must be intelligent life out there?

It's all down to something called the Fermi Paradox, which states that:

1. The Sun is just like any other star, and there are billions of other stars in the galaxy, many of which are billions of years older than ours.

2. Some of these Sun-like stars will have Earth-like planets orbiting them. Any planet with certain conditions similar to ours (such as being an appropriate distance from the star it orbits) could have evolved intelligent life.

3. Given that some of these planets are billions of years older than ours, some of this intelligent life will probably have developed abilities beyond our own, communicating with and travelling to different solar systems and galaxies, something we are now trying to do ourselves.

Conclusion: there are a huge number of other suns out there, with Earth-like planets in their solar system, so intelligent life capable of reaching our planet must have evolved somewhere. Why then have they not visited us yet? Maybe they have, just before we evolved into our current form? When will they be back? Who knows!

---

* Kepler-186f was discovered in 2015 and is the first Earth-sized planet we know of that orbits a star (a bit like our Sun) at a close enough distance to allow liquid water to pool on its surface (aka the 'habitable zone'). Other planets discovered in the habitable zone are usually dissimilar in size to Earth, meaning that any life on Kepler is far more likely to resemble that on our own planet.

While he's not suggesting every planet that resembles ours will have life on it, Prof. Morris thinks that, *should* we find intelligent life, it would be likely to have similar features to humans. And he doesn't stop there: his theory extends to flowers, fish, trees, birds, mammals ... essentially, any life that has developed under a similar set of conditions to those found on Earth, and occupies all the different ecological niches available (land, sea, air), is going to end up looking a lot like it does on good ole planet Earth. Such is the power of convergent evolution.

Interesting theory, although, if we spent billions of pounds, as well as a lifetime's worth of work, on the quest to find E.T., only to finally emerge from a spaceship to be met with a scene not too dissimilar from the one on our own humble planet, we'd be a tad disappointed. We'll take the dreams of Steven Spielberg and little green men any day.

CHIC FACT: Astronomers Jill Tarter and Margaret Turnbull put together a list of 17,129 nearby stars most likely to have planets that could support complex intelligent life. Top of their list was the slightly orange Epsilon Indi A, a star just 11.8 light years away, in a corner of our Milky Way.

While exploration of galaxies to find intelligent life is all well and good, we still have so much more to discover about our immediate neighbours – the planets, dwarf planets and moons that make up our solar system. And none seems to have piqued our interest more than Mars.

Life on Mars (and Beyond)

> He's in the best-selling show
> Is there life on Mars?
>
> *'Life on Mars', David Bowie*

Until recently one planet gave great hope to science fiction writers and astronomers alike that intelligent life may exist upon its crusty surface, so much so that a now very familiar term for its potential inhabitants was coined.

Our second closest planetary neighbour, that beautiful rusty-red orb, has captured our imaginations ever since the second century BC when the ancient Egyptians started to make their meticulous observations. In the seventeenth century German priest and scientist Athanasius Kircher wondered if we could ever make contact with it. In 1996, film director Tim Burton amused us with his imaginings of what might happen if its inhabitants attacked. It will no doubt continue to inspire artists and scientists alike for centuries to come. We are, of course, talking about Mars and the possibility, as David Bowie once sang, that there could be life on it.

When NASA finally announced they were going to drop a car-sized rover explorer on its surface we'll admit to being a tiny bit worried that the lander might get blasted by some irritable troupe of Martians with elongated foreheads and beady eyes on stalks, armed with very powerful laser weapons; or, on second thoughts, in the light of the previous part, perhaps looking just like us.

As it turns out, we needn't have lost sleep over it. The first two rovers, named *Spirit* and *Opportunity*, landed in 2003, followed by *Curiosity* in 2012, and they have been eagerly beavering away on the surface of Mars, undisturbed by aggressive natives, ever since. This doesn't mean that Mars does not harbour life; oh no. We've just had to revise our expectations down from the intelligent life of late twentieth-century film and television, setting our hopes instead on alien micro-organisms. At the time of writing no definitive evidence of life on Mars has been identified, but it seems to be simply a matter of time given the rich bounty of staggering new insights into the planet's composition, along with its history and likely future.

So What Do We Now Know About Mars?

Mars Past

It turns out that in the distant past, Mars wasn't too different from present-day Earth. It had snow-capped peaks, clouds over warm, salty seas and even freshwater lakes. But somewhere down the line something distinctly unpleasant happened that caused it to lose most of its surface water. We're not sure exactly what. We're just glad we weren't around to witness it. It must have been a pretty heavy encounter to leave it in the sorry state we find it in today. That said, many moons ago Mars offered conditions that would have been potentially hospitable to life for several millennia so it may well have evolved at some point along the way, in which case there's every chance that some little microscopic beasties could still reside there to this very day.

Mars Present

The first major revelation from the *Curiosity* data that knocked us for six was the fact that there seems to be liquid water flowing on Mars today. Imaging machinery managed to pinpoint signs of hydrated minerals on slopes in the areas of Mars where streaks are seen – streaks that were dark in warmer seasons, but faded in cooler ones – suspiciously like the transition between liquid water and ice as the seasons change. And if our planet is anything to judge by, where there's water there's usually life.

As for carbon-based organic compounds, the search continues. While they've found six different organic compounds on the planet it isn't completely clear where these actually came from.

The third and final revelation was that Mars is constantly bombarded by radiation from galactic cosmic rays and solar eruptions. Unfortunately, this means that any life on the surface, even in microbial form, would be likely to be wiped out and only that deep beneath the surface will have survived.

> **GEEK CORNER:** One of Mars's moons, Phobos, has unsightly stretch marks around its belly, and these are ones that won't be helped by cocoa butter. The stretch marks are evidence that the gravitational pull of Mars is causing it to fracture as colossal tidal forces push and pull to form grooves in its outer crust. On top of that, it's destined to crash into Mars's surface in the not so distant future anyway, though the brutal gravitational pull will spell its demise long before it strikes the Martian surface. Poor double-doomed Phobos. In the meantime, as far as NASA is concerned, as the larger of Mars's two moons, Phobos is a serious candidate for a base from which Mars landings could be coordinated.

Mars Future

If the film *The Martian* left you with the impression that NASA are pretty determined to get to Mars as quickly as possible, you'd be right. Having won the race to the Moon and with the Earth looking increasingly screwed up by mankind's activities, NASA is now pouring cash into preparations for getting man to Mars by 2050. It seems there's a distinct possibility that at some point in our lifetime a hardy bunch of daredevil adventurers will attempt the seven-month crossing to become the first humans on moist Martian soil.

As cool as that would be, we're probably not going to sign ourselves up because we really didn't like the look of the terrible sandstorms that ripped through base camp in *The Martian*. Sorry Matt Damon, but we're going to take our chances with planet Earth instead, at least for the time being.

Where Else Might There Be Life?

It turns out that Mars isn't the only place in our solar system that you can find a little bit of the old life-supporting liquid H_2O.

For example, 400 million miles away the moon Europa is pushed and pulled so fiercely by Jupiter's hugely powerful gravitational field that, despite being so far from the Sun, the resulting seismic events are likely to produce enough heat to keep water in liquid form beneath its icy surface. A salt ocean has recently been spotted on the surface of Ganymede, another of Jupiter's moons (with sixty-two of them, Jupiter has plenty to choose from).

CHIC FACT: Jupiter is a giant planet, composed mainly of hydrogen and helium; it's so large that the Earth could fit inside it a thousand times. However, it spins so quickly that a day on Jupiter goes by in just ten hours!

Yet another promising candidate is one of Saturn's moons, Enceladus, boasting a terrain of hot sandy springs. In fact, our celestial neighbourhood is so much soggier than we previously thought that senior scientists at NASA have revised their estimates to suggest that we could detect life beyond Earth in the next ten to twenty years. It may not be on the planet Mr Bowie pondered about, but life on other planets and moons may well be out there.

The president of the European Space Agency, Professor Johann-Dietrich Woerner, thinks NASA's ambition to put man on Mars is a little premature. Traversing the millions of miles it would take to get to Mars and back again poses all sorts of problems. So Prof. Johann, being something of a pragmatist, has set his sights on the Moon instead, with hopes of colonising it in the next fifty years.

The Dark Side of the Moon

Imagine waking up, hitting snooze a few times, then finally hauling your lazy arse out of bed to draw the curtains, only to be confronted by the awesome sight of planet Earth hovering over the horizon. Not a bad view first thing in the morning, and, given that the Earth is about four times the size of the Moon, it would probably seem much more impressive, not to mention more colourful, than our well-worn view of the Moon.

All sounds a little bit sci-fi, doesn't it? But with the European Space

Agency (ESA) seriously setting their sights on colonising the Moon in the next fifty years, we thought for a while that it could one day become a beautiful reality for some lucky space-farers.

GEEK CORNER: We humans first set foot on the Moon around fifty years ago but still only a dozen people have felt the crunch of Moon dust underfoot. Since 1972 no one has actually bothered to go back. So why the plans to colonise it now? Inter-agency collaboration and young dynamic new space companies mean the costs of doing so have been slashed from an eye-watering $100 billion to a more reasonable $10 billion.

The only trouble with this dreamy scenario is that if you *were* actually chosen to be one of the first people to take up residence on the Moon, all you would ever see when you drew back those curtains is perpetual darkness. The reason for this rather irksome situation is that the current ESA plans involved building habitations exclusively on the far side of the Moon, aka the Dark Side.

CHIC FACT: The *Dr Who* episode 'Frontier in Space' imagines a penal colony on the Moon by 2540. If the ESA plans go ahead we could have an intergalactic version of old Australia a few hundred years earlier than that.

Now, we don't want to start telling space experts how to do their jobs or anything, but if it was up to us we'd build the colony on the Light Side of the Moon. Why? Partly because we want a nice view of the Earth, but, much more importantly, so that the chosen Moon-dwellers would be able to get enough sunlight to enable their bodies to manufacture vitamin D.

Vitamin D is vital to our bodies and in particular our brains, where it is used to create and maintain all sorts of important neuro-transmitters, helping our brains to function properly. And vitamin

D supplements in tablet form can mess with your immune systems so that's not a great alternative source. Try telling that to a bunch of astronomers hell-bent on taking advantage of the complete absence of light pollution on the Dark Side, which, even in the most remote places on Earth, causes havoc with the more sensitive telescopic data.

We could try to convince them that a fair compromise would be to live on the Light Side, but near the border of the day/night line – which never changes because the Moon always presents the same side to the Earth – and simply hopping on a moon buggy for the short ride to the telescopes over on the Dark Side of the line. Our guess is that they probably wouldn't listen.

What will we be left with if they get their own way? Probably some very pale star-gazers, with incredible intergalactic insight and knowledge, but minds which don't quite function like the rest of ours. Sounds a little like a band we once knew who penned a certain album named *The Dark Side of the Moon*.

> **GEEK CORNER:** The discovery of H_2O on the Moon in 2009 got everyone excited. Not only could it be used as a possible source of drinking water, but it could potentially be split into hydrogen gas used in rocket fuel and oxygen for that essential human function of breathing.

All these incredible discoveries about outer space are only made possible thanks to the breakthroughs in technology that have made space travel and exploration a reality. Groundbreaking innovations such as the following:

Starlight-Propelled Spaceships

One of the most important things to consider when dreaming of exploring our infinite universe and discovering what wonders it holds is how to get around it. And, weirdly enough, one of the most exciting prospects for intergalactic travel today has a surprisingly close relationship with the humble pencil.

Indulge us if you will for a moment. Grab a standard pencil (HB is best) and scribble on a piece of paper. Now take a bit of sticky tape and place it sticky side down over your scribble. Peel the tape slowly back off the page and look very closely at the part of the tape where you've lifted some of the carbon from the pencil markings. You should see tiny layers of compressed graphite attached. If you happen to have a light microscope to hand (available in all good toyshops!) you may even be able to see patches where this layer of graphite is incredibly thin – a single atom, in fact. Strictly speaking it's impossible to see down to the atomic level with a light microscope – although still worth checking out because it's a very pretty sight to behold – but whether viewing it through a microscope or with the naked eye, what you have before you is potentially one of the most incredible of recent discoveries. Behold! Graphene, the twenty-first century's wonder material.

THE GRAPHENE EQUATION

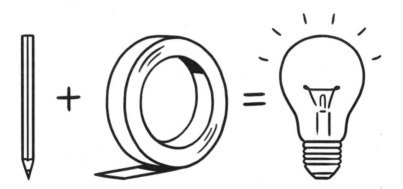

Graphene, like a diamond, is pure carbon. The difference is in the arrangement of its atoms, because while the carbon atoms in a diamond are organised into a 3D crystal lattice, in graphene they are arranged in a flat sheet just one atom thick, resembling a series of honeycombs. This structure seems to give graphene all sorts of miracle properties which, to be honest, science is really only just starting to get to grips with. It's super-light, flexible and incredibly thin, but also tough – two hundred times stronger than steel! It conducts heat and electricity very well, but also acts as a barrier stopping almost anything passing through it, even helium – a ridiculously tiny atom comprising just two protons and two electrons.

It's this vast array of desirable qualities that give graphene its superpowers. From water purification to mobile phones, to electric cars and battery tech, this wonder material looks set to revolutionise all sorts of industries, and even Iron Man is on record as stating that he'd like his next suit to be fashioned from graphene.

The only thing is, like Superman in the presence of kryptonite, graphene seems to lose all its superpowers the minute you try to scale it up by, for example, stacking it into layers. Retaining all these fantastic properties on a larger scale still remains a considerable battle more than a decade of experimentation on from its initial discovery.

What exactly has all this to do with starlight-propelled spacecraft? Well, while trying to figure out how to hang onto all the desirable capabilities of graphene when it's more than one pitiful atom thick, some bright spark decided to make a graphene sponge. The sponge – which looked a bit like a burnt marshmallow – was made by fusing together a load of crumpled sheets of graphene oxide. It turned out to be light enough to sit atop a soap bubble without it popping, as well as super-absorbent – handy for the next time an oil company decides to drill somewhere leaky.

GEEK CORNER: Graphene sponges are essentially crumpled heaps of carbon atoms formed by assembling them in a solution of alcohol and water placed under very high pressure, followed by a process of freeze-drying, followed by heating to create the surprisingly rubbery end product. If you're tempted to have a go at producing some of this wonder material in your very own kitchen, we have some advice for you. Don't bother. The high pressures required are impossible to produce without specialist equipment and even if by some miracle you did manage it, there's a high likelihood it would blow up in your face – so really: please don't try this at home!

So far, so good, but here's where it starts to get really exciting. A team of scientists noticed something truly incredible when they tried to cut the graphene sponge into different shapes with a laser – it moved! Not just to a miniscule degree only observable with a microscope, as had been observed previously under different circumstances. This time it actually shot a whopping 40 centimetres into the air.

After verifying that, no, a tiny Derren Brown hadn't been messing with their minds and neither had a poltergeist been at play, the team set their minds to working out what was going on. Could it be that the laser was superheating the air near their lightweight yet large graphene sponge, causing the propulsion that way? The team put this theory to the test, repeating the experiment in a vacuum where, by definition, all the air had been removed. The same thing happened: the sponge shot into the air and séance frauds around the world turned green with envy.

With the superheating theory disproved, a new one was put forward – that graphene sponge absorbs energy from the laser and builds up a charge, until it's so full of electrons that some effectively burst out and cause the sponge to move. This one seems reasonable; however, no one could really understand why the electrons would only be ejected in one direction, as they could equally have been spat out in all directions simultaneously and the sponge would have stayed put. But then, if we had all the answers, this story wouldn't be half as interesting, would it?

CHIC CORNER: Graphene joins the ever-growing list of important scientific discoveries made completely by accident, including penicillin, pacemakers, Viagra and Vaseline.

What next? Well, the team wondered if anything other than laser light would do the trick. A Xenon light bulb, which emits radiation not dissimilar to sunlight, worked just as well and the sponge's movement was again visible to the naked eye. For their next feat of brilliance they managed to get the graphene sponge to move simply by focusing sunlight onto it with a lens.

It's all still early days and the starlight-propelled spacecraft may be a long way off, but the more geeky half of this science-broadcasting and writing duo (Dr Jack) got particularly excited about this discovery. Why? Because history has demonstrated over and over again that the greatest discoveries in science usually don't start with a *Eureka!* but instead with a *hmmm . . . that's odd, I wonder why that happened?* In other words, the greatest discoveries are often made by accident, when the original aim of the experiment was completely different. That seems to be the state of affairs here with our pencil-loving friends.

Only time will tell whether Dr Jack's spidey sense for an amazing, world-changing discovery is on the money or way off the mark, but, if it does eventually happen, it could revolutionise space travel. Why? Because one of the biggest problems space travellers have is running out of fuel, so having a super-light spacecraft which runs on starlight alone, readily available at all your local galaxy superstores, would be a godsend, or more like a universe-send.

While the geek half of the duo dreams about the limitless possibilities of future space travel, the chic half (Lliana) loves it for a whole different reason: because there's something a little bit magic about looking up at the sky at night, knowing that the very stars up there could one day also help us to discover them.

While the exciting efforts to revolutionise space travel continue, so too does the exploration of other planets in our solar system, including a certain newbie . . .

Pluto's Nemesis

Poor old Pluto. He was demoted to dwarf planet status in 2003 when Dr Mike Brown spotted a similar-sized lump of rock orbiting the Sun (which he later named Eris) and astronomers figured that there were probably quite a lot of as yet undiscovered Pluto-sized bodies in our solar system. Rather than make a mockery of planetary status by

adding more and more oversized asteroids into the club (and confusing the poor schoolchildren of the future), the scientific community decided to raise the bar of acceptance into the Planetarium and make the threshold more exclusive. Anything Pluto/Eris-sized should be referred to as a 'dwarf planet' henceforth.

> **GEEK CORNER:** Everything we've ever seen in the universe only accounts for 5 per cent of its total mass. Absolutely maddening. Where is all the rest? Twenty-seven per cent of the universe's overall mass is dark matter and the remaining whopping 68 per cent is dark energy. Yet neither of these has ever been detected, despite a couple of decades' worth of hunting.

And just in case being demoted wasn't embarrassing enough, Pluto then had to suffer the indignity of being *replaced* in 2016 when the very same Dr Mike Brown identified a new ninth planet – dammit, Dr Mike, what did Pluto ever do to you? The new kid on the block is the soon-to-be-renamed Planet X*. At around ten times the size of planet Earth it should have no trouble gaining acceptance into the planetary family whatsoever.

---

* A personal plea to whoever gets to name the new Planet X: on behalf of us, and all those like us who spent hours upon hours learning the order of the planets in the solar system by endlessly chanting 'My Very Educated Mother Just Showed Us Nine Planets' (for Mercury Venus Earth Mars Jupiter Saturn Neptune Pluto), please, PLEASE, for the love of all things galactic, start the new planet with the letter P.

CHIC FACT: A new joint collaboration on a probe called Wukong (between researchers in China, Switzerland and Italy) has been launched into space packed with a load of apparatus, including dark matter detectors, so hopefully we won't have to wait too much longer for definitive proof that it exists. Oh, and Wukong is an apt name for such a device – 'Wu' roughly translates as 'understanding' and 'Kong' means 'space', so assuming all goes well it should hopefully live up to its name. Wukong also happens to be the name of the Monkey King in the classic Chinese legend *Journey to the West*, so given that it's an international collaboration this also seems to bode well for the project's feng shui.

Here's the catch – no one has actually ever *seen* Planet X, but astronomers believe they have the evidence to prove it exists. Despite having never been directly observed, they do know with some certainty that it must be 15,000 miles wide (making it a Neptune-sized planet) and that it orbits our Sun once every 15,000 years. It's all down to a strange clustering that they observed of six other previously known lumps of rock that orbit beyond Neptune. Using a series of mathematical and computational models they figured out that the only thing that could possibly have gravitationally herded those six heavenly bodies into their weird orbits is Planet X. The chances of these six orbits being the shape they are without a Neptune-sized lump of rock swinging into their midst every 15,000 years is infinitesimally small.* Mike Brown's team predict we will be able to see it via telescope in the next five years, but for now, just like dark matter, dark energy and all the other currently invisible entities in the universe, we are going to have to give way under the weight of all the evidence and accept that it *has to exist*. Sorry, Pluto, but it looks like Planet X is here to stay.

---

* There's only a 0.007 per cent chance (1 in 15,000 – what is it with that number?) that the 'real' ninth planet, aka Planet X, doesn't exist.

> **CHIC FACT:** Looking for a quick way to drop the pounds? Weigh yourself on Pluto – a 15-stone person would only be around 10 stone on the dwarf planet thanks to its lesser gravitational pull. Go to Jupiter on the other hand and your weight would almost double.

Pluto may have been demoted to 'dwarf planet' but it wasn't always so! In fact, the story of Pluto is a rich and interesting one.

Pluto's Tale

It sounds almost too good to be true: a discovery that would make Walt Disney smile from beyond the grave, but the news is in and scientifically verified: Pluto does have a tail. So, in the spirit of Mr Disney, let us tell you a little story.

Once upon a time, back in 1906, a baby named Clyde Tombaugh was born to an Illinois farming family. The child grew into a smart young boy, with ambitions to go to college to study astronomy. One fateful day a catastrophic hailstorm occurred, destroying all the families' crops. Their fortunes dwindled in the aftermath and with them went the boy's dreams of a university education. Undeterred by this obstacle to his academic aspirations, the boy grew into a young man who tinkered with lenses and mirrors to build himself a variety of telescopes, which he would point at the sky, night after night after night.

Young Clyde sent sketches he made of what he saw down the barrel of his home-made instruments to a magical place known as the Lowell Observatory in Flagstaff, Arizona. So impressed were they by his work that they promptly offered him his dream job. Clyde would spend the next sixteen years exploring the wonders of the universe. He would unearth all sorts of wonderful things, including his best-known discovery of all: Pluto.

Then, one day in 1992, a now much older Clyde was contacted by a man named Robert Staehl from the Johns Hopkins Laboratory, who asked his permission to visit Pluto. Plans were being drawn up for a 'Pluto Flyby', the first mission to fly a spacecraft directly past

Tombaugh's discovery, in the hope of capturing some of the secrets of the mysterious distant planet. The probe, named *New Horizons*, took off in 2006, and nine and a half years later, on 14 July 2015, it finally reached its target, snapping all sorts of incredible images of Pluto and its five moons as it flew past. Thereafter it careened into the depths of outer space, where it will continue its adventure for evermore. And while much of the information captured by *New Horizons* will take years to be beamed back to planet Earth, early data provided the world with some fascinating insights, such as the incredible fact that Pluto has a tail, created by solar winds sweeping its nitrogenous atmosphere back into a 50,000-mile-long, densely packed sweep of ionised gas and plasma. Pluto also has an icy heart (like that of a wicked stepmother in a Disney film): a large area of its surface that looks white and heart-shaped in photos. This region, which is actually a frozen carbon monoxide ice lake, has been named by astronomers Tombaugh Region.

Sadly, our old friend Clyde never got to find out that his beloved Pluto had a tail or a heart, as he died before the mission set sail across the solar system. But this did mean he could become an intrinsic part of the project. His ashes were launched with the probe, and so the very man who discovered Pluto ultimately joined the flyby and even now continues his life's calling, pushing on into the unfathomable depths of deep space. A fitting Disney ending to a different kind of tail, the tale of Clyde Tombaugh.

GEEK CORNER: Along with Clyde Tombaugh's ashes there are actually eight other items hidden aboard the *New Horizons* probe: 434,000 Earth people's names on a CD-ROM; a 25 cent (quarter) coin from the state of Florida, where it launched from; a Maryland state quarter, where *New Horizons* was built; a small piece of *Spaceship One* (the first privately funded manned spacecraft); a CD-ROM with photos of all the people who worked on the mission; two versions of the American flag; and a 1991 US stamp with the words 'Pluto: Not Yet Explored' printed on it.

If you think knocking Clyde Tombaugh's beloved Pluto off the list, only to put a much more massive hunk of rock back up there in ninth place in our solar system's planetary race is harsh, imagine the uproar if the Big Bang Theory was toppled from its pedestal!

Somewhere over the Rainbow Universe

Picture this: you're driving a quad bike around an athletics track. Your task is to go as fast as you can without coming off the track, or, worse still, rolling your chunky-wheeled steed by being a bit too liberal with the throttle. Sounds fun, right? Now imagine someone else being given a quad bike and invited to do the same thing, but in the opposite direction. Once you've had the chance to do a few laps to reach top speed on the straight, you are now instructed to aim for each other, head on, no touching the brake ... Not so fun, right?

This head-on collision is roughly what happens when tiny pieces of atom are accelerated in opposite directions by a 16.5-mile loop of hugely powerful electromagnets in a particle accelerator. The pieces of atom achieve such incredibly high speeds that when they crash together the subatomic particles smash into even smaller smithereens. Why bother? Well, from these teeny, tiny elementary acorns everything in our unfathomably large oak-like universe is formed. Understanding the nature of the universe is the name of our game.

CHIC FACT: Some people reckon that when Obi-Wan Kenobi said in Star Wars that the Force 'surrounds us and penetrates us; it binds the galaxy together' he could have been really talking about gravity!

We are, of course, talking about what happens inside the feather in the European physics community's cap that is the Large Hadron Collider. One hundred metres beneath the Swiss/French Alps, this 100,000-ton monstrosity was built at mind-boggling expense. Between 2014 and 2015 the old dog* was given a breather from all this incessant subatomic particle bashing for a spring clean and an upgrade. And here's where things get really interesting, because when they fired the not-so-little beauty back up again in March 2015 witnesses saw, through the haze of what they at first presumed to be the smoke of a short circuit, 'a shimmering halo that spanned the full spectrum of visible light' that promptly disappeared after 2.6 seconds. What they had accidentally discovered is now suspected to be none other than enigmatic, much discussed but never observed, RAINBOW GRAVITY.

GEEK CORNER: If what they saw on that fateful day in March 2015 at the Large Hadron Collider really *was* rainbow gravity then we might finally have something we can use to unite the physics of the big stuff (relativity) with the physics of the tiny stuff (quantum). To reach those heady heights of theoretical unification we're going to need a lot more evidence than a shimmer of coloured light in a puff of smoke. We can either patiently wait for an immensely energetic natural event – ideally a gargantuan supernova – with enough oomph for the rainbow gravity to show up on a grand and undeniable scale or upgrade the world's telescopes so we can see it in everyday supernova. On the pitiful timescale of the average human lifetime gargantuan supernovae are quite sporadic so don't hold your breath. That said, it seems that the Large Hadron Collider's fortuitous mishap may have given the global astronomy community the kick up the bum to splash out the dosh needed to make their devices rainbow gravity sensitive.

---

* The old dog – the Large Hadron Collider – was built during the decade between 1998 and 2008 and started to conduct its first experiments in March 2010.

What Exactly is Rainbow Gravity?

Music lovers among you would be forgiven if you thought it was something The Klaxons sang about in their Mercury Award-winning album. Meanwhile, literary lovers may hark back to Thomas Pynchon's novel *Gravity's Rainbow*. However, for the science bods among us, rainbow gravity is actually the theory that as light travels through space, all the different wavelengths (think colours of the rainbow) are affected by gravity in different ways. Infrared is curved more than red light, which is itself curved ever so slightly more than green light; green more than blue light; blue more than ultraviolet; UV more than X-rays and so on.

GEEK CORNER: If what they saw on that fateful day in March 2015 at the Large Hadron Collider really *was* rainbow gravity then we might finally have something we can use to unite the physics of the big stuff (relativity) with the physics of the tiny stuff (quantum). To reach those heady heights of theoretical unification we're going to need a lot more evidence than a shimmer of coloured light in a puff of smoke. We can either patiently wait for an immensely energetic natural event – ideally a gargantuan supernova – with enough oomph for the rainbow gravity to show up on a grand and undeniable scale or upgrade the world's telescopes so we can see it in everyday supernova. On the pitiful timescale of the average human lifetime gargantuan supernovae are quite sporadic so don't hold your breath. That said, it seems that the Large Hadron Collider's fortuitous mishap may have given the global astronomy community the kick up the bum to splash the dosh needed to make their devices rainbow gravity sensitive.

What's This Got to do With the Big Bang Theory?

Einstein's famous theory of relativity explains that all things in the universe are affected by gravity *equally*. And relativity is the essential framework for the entire Big Bang Theory, so, if you find hard evidence of rainbow gravity (contradicting the idea that all things are affected by gravity in the same way), then that casts doubt on the very foundations of the Big Bang Theory.

If that happens, and the Big Bang is filed away into the drawer labelled 'Good-But-Wrong-Ideas' (alongside shoes for dogs and crotchless underwear), then it looks like the universe has lasted forever and always will. We're not gonna lie, that suits us very well indeed. We always found the idea of there being a boundless nothingness before the Big Bang came along and created absolutely everything a little bit hard to fathom. An infinite forever sits much more comfortably.

As our minds grapple with the idea that the Big Bang Theory could be debunked, the quest is on to understand where the building blocks of life came from. Something that may hold some of the secrets to the origins of life are comets, formed at the same time as planets and stars, as long as 13 billion years ago. Finally, we can get up close and personal . . .

Catching Comets

Some people believe that if you see a shooting star and make a wish it will come true, but where did this idea originally come from? It may well have stemmed from the second century AD, when Greek astronomer Ptolemy wrote that as the gods peered down onto Earth between the spheres, stars would fall between the cracks and tumble through the night sky. He reasoned a shooting star was our only clue that the gods were looking down upon us – a prime opportunity, he said, to have your wish heard by the very immortals who could make them come true.

CHIC FACT: Swabians (native to southwest Germany) believed that seeing one shooting star meant you'd have a year of good luck. See three, however, and you were doomed to die! In the Phillipines if you see a shooting star you're supposed to tie a knot in your hankerchief before it disappears for good luck, whereas in Chile you have to pick up a stone.

Whether or not you believe such superstitions (we are *still* waiting for the all-expenses paid, round-the-world trip we wished for) one thing is for sure – there is little more magical than seeing the sky at night lit up by celestial streaks. So what exactly creates them?

Usually it's a meteor – particles of rocks and dust that fly off comets and burn up as they reach the Earth's atmosphere, creating 'shooting stars'. The comets themselves are leftover chunks from when the planets and stars were forming, and leave beautiful tails of gas and dust in their wake. Comets are responsible for some of the most magical and mystical sights in the sky, but how much do we really know about them?

Despite it being likely that there are billions of comets just beyond our solar system we know surprisingly little about them. Why? Because they move too damn fast for us to get close enough to analyse them properly ... that was until the discovery of lazily named comet 67P, which is 305 million miles away zooming through space at breakneck speeds of up to 24,600 mph.

In February 2004 the European Space Agency decided to try to catch up with 67P to find out more, in the hope that the wee fella might unlock some of the secrets about how our own planet, and indeed all life, began.

And so, we blasted a rocket into space carrying a spacecraft called *Rosetta*, which then went pinballing from planet to planet over the course of ten years like a drunken Scotsman dancing the Highland fling.

As chaotic as the path traced by *Rosetta* may look to the untrained eye, it was in fact sent on a precisely timed, carefully plotted trajectory that enabled it to catch up with the comet ten years later, at exactly the right speed to fall into the catcher's mitt that is the ten billion-ton lump of ancient rock's weak gravitation pull. We don't know about you, but we find it absolutely mindboggling that it's possible to successfully plot a 3,650 day trip that covers almost four billion miles with enough precision to not just catch up with a comet but to fall into orbit around it. Not so much a case of catch a falling star as catch up with a falling *comet*.

CHIC FACT: Comet 67P is so far away that it takes seventeen minutes for every bit of information sent from it to travel back to Earth.

That's only part one. Once the team managed somehow to get *Rosetta* into orbit around 67P they dropped a washing machine-sized robot called Philae to the surface from fourteen miles above the aforementioned, rubber-duck-shaped, rocky mass. They hoped that by aiming for a specific spot on the comet that would leave Philae basking in the sunshine, it would be able to harvest energy with its solar wings to supply plentiful juice (aka electricity) to all of the scientific gear on board, including drills and elaborate chemistry sets.

There we all were, watching from planet Earth with bated breath, ten years after *Rosetta* was launched, as Philae dropped down onto 67P and the world was all set to proclaim the team behind the mission heroes. Cue the champagne, right?

Sadly, as in all good sci-fi plots, there was a twist, and in this case life is no stranger than fiction, because critical to the success of this wild dream was for Philae to land without bouncing. Unfortunately, the harpoons that were supposed to deploy upon touchdown to anchor Philae to the ground failed to make an appearance, and so bounce it did! What is it they say about the best-laid plans?

The watching world took a collective sharp intake of breath, and fortunately didn't hold it; despite bouncing about a kilometre into the air and threatening to disappear off into space entirely, it wasn't quite high enough for Philae to rip loose from 67P's gently seductive pull altogether. And so she did come down again. Eventually. A full hour and fifty minutes later. The world cheered, whooped, hugged and tweeted, but what next?

After one more bounce – this time a much more modest seven-minute 'hang time' – our poor little robot found itself lodged at the foot of some kind of cliff or rocky outcrop, which unfortunately blocked the Sun from its solar sails. It managed to do a bit of 'science' – fifty-seven hours' worth – drilling for soil samples and sending back data about what the outer layers of comet dust, collected at different altitudes on the way down, are made of. But after a meagre transmission of just a few photos of the descent, and a couple of sputtering messages later, it conked out with flat batteries. Not quite the full month the team had hoped she would survive for.

However, all was not lost: after seven months Philae sputtered back to life. During its slumbers it had travelled a whole lot closer to the Sun and found itself approaching the comet's perihelion – the location in space where it comes closest to the Sun – 67P summertime, if you like. Closer to the Sun means more energy hitting the solar panels and finally charging the batteries enough for it to transmit a single eighty-five-second message one Saturday, and three tantalisingly brief messages the next day, each of them lasting just ten seconds.

GEEK CORNER: The *Rosetta* spacecraft was named after the famous Rosetta Stone, currently residing in the British Museum, London. Its discovery in 1799 enabled academics to finally decipher Egyptian hieroglyphics after hundreds of years of scratching their heads wondering what all those funny engravings were banging on about. How did a 762-kilogram lump of rock enable this revelation? Because the same message was inscribed on it twice: once in ancient Egyptian and again in ancient Greek. The inspiration for naming Philae is similar but less well known. It is the name of an island in the River Nile on which another obelisk was discovered that also had a bilingual inscription in both ancient Egyptian and ancient Greek, including the names of Cleopatra and Ptolemy.

Despite the incredible achievement that mankind pulled off – landing a robot on the face of a comet for the first time in history – was it really worth it? And what did we actually learn? Well, for starters, from the sound of Philae's bounce we learned that comets are much harder than we suspected, meaning they're probably made of solid ice beneath a layer of dust. (Not bad.) We also discovered there was water on 67P, although a different type of water from ours, meaning comets are unlikely to be the source of water on our planet – one of the essential building blocks of life. (OK, getting better.) And we discovered molecular oxygen, something no one really expected, as it had never been found on a comet before. (Oooh-errr, now we're impressed.)

However, most excitingly of all was that Philae discovered organic molecules, which means that if a comet smashed into a planet with the right life-enhancing conditions, like water and warmth, it's entirely possible that comet's impact can promote the evolution of life. (Now we're talking!) From creating shooting stars, to holding the potential to create alien life: comets, you beauteous things, we salute you.

It's impossible to talk about space travel without mentioning the most beloved of TV series, Star Trek. *But has science caught up with science fiction yet? One recent breakthrough suggests it might have.*

'We're Caught in Her Tractor Beam, Captain'

tractor beam: [noun]
[in science fiction] a beam of energy that can be used to move objects such as spaceships or hold them stationary.

Oxford English Dictionary

Famously ill-tempered chief engineer Scotty is *Star Trek*'s equivalent of Gordon Ramsay: prone to explode when things don't meet his punishingly high standards, but at the same time able to pull off the seemingly impossible. But if there's one thing that Chief Engineer Montgomery 'Scotty' Scott couldn't stand, it was getting his beloved vessel trapped in a tractor beam.

Over the many decades that *Star Trek* aired on TVs across the globe, there were countless nail-biting moments when a much larger spaceship got near enough to the Starship *Enterprise* to pincer it in a vice-like grip, rendering everyone on board powerless to escape. The alien race could then hold it there captive, move it closer or release it back into space, toying with it a bit like a cat with a mouse and leaving the crew feeling like the village idiot shackled in the stocks.

Of course, there were no actual mechanical fingers physically grasping the hull of the ship. Instead, a machine projected a force field in a certain direction so that anything caught at the focus of these 'graviton' beams would be left dangling prostrate in space. So far, so sci-fi, but the big question remains – have *Trekky*-loving scientists managed to recreate a tractor beam in the real world?

Until recently the closest anyone had got was to use heat waves from a laser beam to move microscopic particles by a few millimetres here and there, which we're sure you'll agree is comparatively a bit pants. But now, step aside electromagnetic radiation; hello Mr and Mrs Sound Wave!

In 2015 a team of scientists, led by the uber-slick sounding Professor Bruce Drinkwater, developed a panel of mini-loudspeakers that could create intricate, shifting patterns of high-intensity sound waves, probably best imagined as a 3D sound hologram. This flat array of speakers produces a real-life 3D force field which can be moulded into pincers or even mini-tornados, gripping pea-sized beads in mid-air and moving them around in any direction up to thirty centimetres in the air.

It works like this. A combination of ingenious maths and the extremely powerful precision speakers means the ultrasonicians can create small pockets of still or 'dead' air, where the bead sits, surrounded by areas of ultra-high-intensity vibrations. By gradually changing the location of the zone of dead air, while ensuring the high-intensity air particle vibration remains on all sides, they can move small objects about with incredible accuracy.

CHIC FACT: The first ever tractor beam was imagined by E. E. Smith in his 1931 novel *Spacehounds of IPC* – so it only took us the best part of nine decades to make his dream a near-reality.

These so-called tractor beams are essentially nothing more sophisticated than air pressure fluctuations. Of course, you could say the same about all acoustic phenomena, from music to birdcalls, human speech and sonic booms, but none of those are visible to the naked eye. That's what we find most impressive about this breakthrough: you can actually see stuff levitating through the air as if by magic. Sort of like a high-tech hummingbird.

The team even had a little bit of fun with this on a video demo, setting the flat panel of speakers into a cardboard cut-out of a UFO to demonstrate how it can swoop into view and come to a halt, trapping a coloured bead in its invisible sonic tractor beam. The bead could then be gently sucked up into the underside of the faux-spaceship with deft precision.

GEEK CORNER: Another research team had also been working away on these sorts of sonic force field innovations, using an ultrasound device already in use in MRI-guided surgery. This same team also invented something that might just excite fans of another TV series that appeals to multiple generations – a real-life sonic screwdriver.

What's the point of it all? Nice as it is to keep the sci-fi lovers among us amused, there has to be more to it than that ... and there is! For a start, these sonic tractor beams could eventually be used to handle dangerous toxic substances, or even carry out a targeted drug delivery of toxic therapeutic chemicals to precisely where they're needed in our bodies, with no leakage into the surrounding healthy tissues.

Who'd have thought that all those years ago, when Scotty was ranting on about the dreaded tractor beam, it could one day be used to help save the lives of us folk back here on planet Earth?

Final Thoughts

For centuries we've explored our lands and seas, our bodies and minds. But today it's the discoveries we are making in outer space that are the most exciting and revolutionary, extending beyond our own galaxy and way into the universe as a constant reminder of how tiny and perhaps insignificant we really are.

As the brilliant scientific minds of the world continue to make huge advances in technology, innovation and their own travels, we will continue to watch with fascination and awe. Ultimately, despite all our hopes and promises, we may never find the very thing we hope to find – other intelligent life – in our own lifetimes, but with each new discovery on our galactic journeys we also learn a little more about who *we* are and where we came from. And who knows? If we continue to ravage our own planet, perhaps one day we will become the very intergalactic planet-hopping creatures that we are now trying so hard to find.

4

Take a Walk on the Wild Side

Birds do it, bees do it, even educated fleas do it. So let's do it, let's fall in love, get drunk, fart, hold grudges and squeal for joy.

OK, so ascribing our humans traits to animals can be tempting but anthropomorphism is a contentious game. From the chimps that brew wine, through to pandas faking pregnancy and from mice that sing for sex, to feathered art critics, this chapter proves without a shadow of a doubt that the animal kingdom is full of creatures that behave *almost* as weirdly as us human beans.

And we begin with the very first story that inspired *Geek Chic Weird Science*, the podcast . . .

The Panda Who Faked Her Own Pregnancy

It sounds like the kind of headline you might get on episode of *Jeremy Kyle: The Animal Special*. ''Fess up. Panda, you faked your own pregnancy for nice food and posh digs, now it's time to come clean!' – but according to keepers in the Chengdu giant panda breeding centre this sentiment is not too far off the mark.

In 2014, six-year-old Ai Hin was supposed to be the world's first panda to give birth to cubs live on air, but she scuppered the TV company's big plans when it was discovered that she was actually having a 'phantom pregnancy'. Surprisingly, these aren't actually all that uncommon among female giant pandas, who occasionally mimic many of the signs of the genuinely pregnant – losing their appetite, sleeping much more than usual and exhibiting maternal behaviours like shredding bamboo to make nests or tenderly cradling inanimate objects. Sometimes their hormones rise exactly as they would if they were really pregnant!

CHIC FACT: Phantom pregnancies are common in pandas but can also occur with dogs, horses and even humans. The cause is usually psychosomatic, i.e. the brain causing the changes due to a desire to be pregnant. Women can even stop menstruating, have a swollen belly and experience contractions!

However, what was unique in Ai Hin's case was that experts suspected she may have deliberately 'faked it'. You see, panda mothers-to-be at the Chinese centre get swiftly moved into their own private rooms with air conditioning, around-the-clock care and a much nicer menu, of fruit buns, bamboo and fresh fruit. Ai Hin stopped showing signs of pregnancy – such as thickening of the uterus and increased levels of faecal progesterone – soon after she was moved to the VIP digs with better food options. Chengdu panda expert Wu Kongju reckons clever Ai Hin must have noticed the first class treatment expectant pandas received and so mimicked the tell-tale signs of pregnancy to get in on the five star action. We're not sure we blame her – fruit buns are delicious!

Ai Hin isn't the only panda accused of faking it for treats. In March 2015 an eleven-year-old female in Taiwan named Yuan Yuan was artificially inseminated and soon started showing those classic signs of pregnancy. After getting a nice cool, private room and a menu upgrade for over a month, an ultrasound showed that she wasn't pregnant after all. Cheeky!

Pandas aren't the only ones that have tricks up their sleeve to get a better quality of life. There's the Large Blue species of butterfly boasting its beautiful and distinctive blue wings speckled with black dots in adulthood, which mimic the look and smell of the Myrmica ant's offspring during the late caterpillar stage. This fools the unsuspecting ants into taking the baby butterflies into their nest where they can enjoy the twin benefits of extra food *and* protection. And have you ever noticed pet dogs pretending to urinate on demand to earn extra treats? Craftier still, certain species of a bird called the drongo are accomplished impersonators of other animal calls – mimicking everything from glossy starlings to meerkats – to scare away smaller creatures and then steal their food.

Come to think of it, *Jeremy Kyle: The Animal Special* doesn't seem that far-fetched after all. Next week it's: 'Two-Timing Aardvark Takes Paternity Test – The Devastating Truth Finally Revealed.'

CHIC FACT: In a bid to get the libido-challenged bears to have more sex, researchers in China have started showing them panda porn, and it's working! Sixty per cent of their male pandas have sex now, up from a worrying low of 25 per cent. Some keepers even encourage young pandas to have threesomes with older, wiser bears, to help teach them the ways of the world.

While pandas fake their own pregnancies for treats, a smaller, squeakier, bewhiskered little animal is also pulling out the stops to get what it wants, and in this case it's music that provides the key.

The Mice Who Sing for Sex

We've all heard of singing for your supper, but how about singing for sex? Rock stars may have been using it as a seduction technique for decades, but scientists found out that male mice also sing to their partners to get them in the mood for some mousey loving.

Their ultrasonic love songs are in the range of 35 to 125 kHz, far too high-pitched for the human ear – our hearing has an upper limit of 20 kHz. However, despite us humans not being able to join in, the love squeaks are thought to be an important part of mouse courtship – it worked for Elvis and Mick Jagger after all. The male

mouse breaks into a pretty complex song (in terms of tempo and pattern) the moment it smells the female's pee – a normal male reaction to the smell of lady pee, right? He then drops into a softer song, with a simpler pattern, as the female comes into sight and starts to approach him.

CHIC FACT: Mice produce seven to eight litters of between four and sixteen pups a year – and the von Trapp family thought they had the whole family choir thing nailed.

It's been fifty years since this discovery was made and everyone assumed this sexy serenade was a one-way street. However, it was recently discovered that the female mice actually sing back. The reason it took so long for us to catch on is that it's tricky to figure out exactly where a sound's coming from when there's more than one mouse in the picture. However in 2015, researchers in Delaware developed a sophisticated array of microphones and a sound analysis chamber, enabling them to pinpoint exactly where these mousey love songs were coming from. Much to their surprise they discovered that while the male initiated the serenade (remember that alluring, song-inducing pee smell?) the female would join in soon afterwards. Think Sonny and Cher but with a whole lot more fur. She would even slow down as he chased her so that the horny male mouse could catch up as she squeaked, a bit like a mousey Barbara Windsor in a *Carry On* film.

Though all this research into lusty singing mice may seem a little frivolous, the fact that mice can modify the complexity and volume of their love songs depending on whether their potential mate is in sight or not could further our understanding of our own forms of social communication. The study made it clear these weren't just random squeaks – the soon-to-be mouse lovers were actually communicating with each other and coordinating their utterances, depending on their social context, which is something autistic people sometimes struggle to do. So it's hoped that it might help to shed some light on what's going on in autism.

CHIC FACTS: Stephen Spielberg was clearly way ahead of the scientific game. In 1986 his first animated film, *An American Tail*, told the story of Fievel the mouse, and his sister, Tanya, who both sang to their little mice hearts' content.

So exciting is this burgeoning area of research that scientists all over the globe are now studying the vocal abilities of mice, and uploading the squeaky warblings onto MouseTube, storing the ultra-sonic vocalisations of many a rodent. Sadly, Geek Chic's proposed album, *Mickey & Minnie's Classic Love Duets*, will *not* be coming to a shelf near you anytime soon.

Thought mice (and humans!) were the only ones partial to a bit of musical revelry? Think again; zoomusicology – the study of animal music and sound communications – is a field in and of itself. All sorts of creatures use sounds and music to communicate, defend their territory or woo potential lovers, including dolphins, bats, seals, whales and . . . marmosets.

Musical Marmosets

This next topic comes with a serious cuteness-overload warning, because it's been discovered that marmosets – tiny South American monkeys small enough to fit in the palm your hand – have genuine musical abilities. Perhaps they could form a little marmoset orchestra, with tiny marmoset recorders, marmoset violin players, and a marmoset conductor in a minuscule tux . . . ?

OK, OK – back to the science!

Their abilities largely centre around what is known as basic pitch perception, which is the ability to tell the difference between, for example, a high-frequency and a low-frequency sound. Can you tell the difference between high-frequency sounds, like shrill whistles, squeals or squeaks, and low-frequency sounds, like thunder or bass drums? Then your pitch perception is intact. Kudos to you.

Although many creatures aside from us humans have the ability to distinguish between high and low notes, no other less-sophisticated mammals were thought to process pitch with quite the same aptitude as us big-brained humans . . . until now.

A team of researchers set out to find out exactly how marmosets perceived pitch. Using traditional conditioning techniques (rewarding the marmosets with food when they got it right) they trained up a bunch of the Brazilian monkeys to lick water from a spout any time they noticed a change in pitch. After training, the marmosets would then be tested by playing them a series of the same repeated note. The moment the pitch changed slightly, out popped a tiny little marmoset tongue to lick the water from the spout. We know, we know, we're veering back into dangerously adorable territory. Come on, science brains: focus, FOCUS!

Armed with the basic knowledge that marmosets *can* detect

small changes in pitch, the scientists took the research further, discovering that the little monkeys were able to perceive pitch in three specialised ways that scientists had previously thought unique to humans:

Firstly, just like us, the marmosets could easily discern tiny changes in sound frequency in the lower octaves, but struggled to detect any difference between similarly spaced tones in the higher octaves.

Secondly, just like us, they were able to pick up on subtle changes in the spread of pitches transitions. If a series of tones increased by 100 Hz and then suddenly by only 90 Hz, they would indicate that they'd noticed it by taking a mini-slurp of water from the spout. We humans take observing these tiny changes in the pattern for granted, but it's actually a very technically sophisticated ability, so it was a huge surprise when it turned out our tiny Brazilian furry friends could also spot such a subtle change.

Finally, just like us, at high frequencies they were able to use rhythm to help tell the difference between different sounds.

But what does this all mean, other than the fact that the miniature marmoset perceives pitch *just like us*?!

Pitch perception is a precursor to speech and musicality, and, seeing as we humans share this ability with marmosets, it follows that we probably also share a common ancestor with similar talents. As the massive prehistoric supercontinent Pangea split up over forty million years ago – at which point the Americas (to which marmosets are native) became separated from every other land mass which humans evolved on – this means that pitch perception must have emerged way back then, far longer ago than previously thought. It's pretty incredible to think that some of our earliest ancestors – the primitive monkeys roaming the Earth such as the genus *Apidium* – had musical abilities way back then in the Eocene period, just twenty-five million years after the extinction of dinosaurs.

Now, about that marmoset orchestra . . .

We've seen that some animals have musical skills like serenading and pitch perception; others, like our cats, simply enjoy listening to it and find a pleas-ant little ditty a great way to unwind. But what kind of music do kitty-cats most enjoy?

Classical Cats

There you are, relaxing in an armchair after a long day's work with your cat curled up cosily on your lap. You reach for the remote to turn on your sound system. But what do you put on? A little bit of Beyoncé? A blast of The Beatles? Or maybe a soothing Beethoven symphony? Before you make your decision, there's one very important question to consider: what would your *cat* like to listen to? Because a 2015 study showed that cats prefer to chill out to classical music.

CHIC FACT: Scientists in Wisconsin have created music specifically for cats, which sounds a lot like a combination of purrs and meows. Cats showed a definite preference for the cat tunes over classical music and the younger the cats the more they seemed to enjoy it.

Scientists in Portugal popped a pair of headphones over twelve cats' ears while they were under anaesthetic (they were being neutered at the time). The unconscious kitties were then played two minutes of either classical, pop or rock music while their pupil size, heartbeats and breathing patterns were monitored. The scientists found that when the sleepy kits were played classical music (Samuel Barber's 'Adagio for Strings', in case you're wondering) they were distinctly more relaxed, as measured by their slower breathing and smaller pupil size. When they were played rock (AC/DC's 'Thunderstruck') by comparison they appeared to be stressed out. Pop music (Natalie Imbruglia's 'Torn') fell somewhere in the middle – so clearly the cats were a bit 'whatevs' about it.

CHIC FACT: Lou Reed once played a gig just for dogs (we're guessing the reviews from the dog critics were pretty one-note).

Some scientists believe that the cats prefer classical music because its tempo matches the frequency of their own purring, while other non-scientists think they dislike AC/DC because they've simply got rubbish taste in music.*

The study did have more significant meaning than just trying to figure out what tunes to put on our pet cat's playlist. The leading veterinary surgeon behind the research was initially inspired by reports that humans feel less pain and are calmer when listening to certain music and wondered if the same could be applied to our feline friends. Turns out he was right and the team reckon that playing classical music while cats (or dogs, or even us humans!) undergo surgery could decrease the amount of anaesthetic needed, increasing the chances of survival as a result. What was it someone once said about a DJ saving their life?

The classical cats study could be rolled out to other animals, like dogs or birds and eventually humans, meaning that survival rates in surgery could improve. However, anyone planning to carry out surgery on sharks may need to dig out a very different collection, because, unlike cats, they seem to prefer something a little heavier.

Sharks Love Heavy Metal

Ever wondered what kind of music sharks like to listen to? No, neither had we – although we had kinda hoped it would be the *Jaws* theme tune – but Australian shark-dive operator Matt Waller found out for himself back in 2011 when he borrowed his friend's underwater speakers and hung them off the back off his boat. He simply hooked up his iPod, pressed play and let the music blast out into the sea.

---

* We may have made up the second part about AC/DC – although we do think cats should do some serious soul-searching on this matter.

GEEK CORNER: Sharks are only attracted to low-frequency pulsing sounds (20–60 cycles/second) – like those found in heavy metal – and not to higher-frequency sounds (400–600 cycles/second). Sorry, Mozart and Kanye, but you won't be making it on to the great white's playlist.

First up alphabetically was AC/DC's 'Back In Black', and before you could say 'Angus Young is the Daddy-o' huge great white sharks started swimming up to the boat. Matt realised he may have hit upon something and started using this technique to regularly attract sharks to his cage-diving tourist trips, finding it worked a whole lot better than chucking chum (mashed up bits of leftover fish) into the sea. Matt also noticed that when he played AC/DC's hits the sharks seemed to be calmer and act less aggressively. They even occasionally rubbed their faces up against the speakers, pretty standard behaviour for any genuine AC/DC fan, we reckon.

CHIC FACT: The sharks' favourite AC/DC tunes were: 'You Shook Me All Night Long' and, quite aptly, 'If You Want Blood (You Got It)'. However, pacific islanders have used this knowledge for many years to attract sharks, but instead of using heavy metal they rattle coconut shells under the water.

A group of documentary film-makers put Matt's metal-loving-sharks-theory to the test once again in 2015 while making a shark show for Discovery Channel, although this time they plumped for a bit of Death Metal. Wanting to lure a huge great white shark (brilliantly named 'Joan of Shark') towards them to get better footage they used military underwater speakers to blast some heavy tunes by death metallers Darkest Hour into the sea. Joan didn't show up, but two others soon did.

But why does our big-toothed frenemy seem to love all things heavy? Sharks, and indeed many fish and cephalopods, have jelly-covered and fluid-filled sensory hairs on their body called neuromasts, which are clever little things*. As a current of water hits them they move in the same direction, exciting the sharks' nerves in a way that lets them know that something is coming, possibly a tasty meal in the form of a fish. The low-frequency pulsing vibrations of heavy music, with all its throbbing bass lines and fast beats, create the same frequency patterns as those of a dying or distressed fish thrashing about. This so-called 'yummy hum' is detected by the shark's lateral line (a sense organ made up of lots of these neuromasts running from their head to their tail) and in their inner ears, letting the sharks know that their food is being served. It seems, then, that heavy metal music acts as nature's dinner bell when it comes to sharks.

CHIC FACT: Sharks aren't the only ones who like heavy metal. One gardening expert genuinely reckons his plants grow faster when he plays them Black Sabbath, whereas blasting some Cliff Richard makes them wilt away and die (no comment).

Sharks may love the sounds of heavy metal, but what about their other fishy friends?

Are Goldfish Classical Connoisseurs?

It's a hard life being a goldfish. They get shoved into plastic bags to be won as prizes at fairs, are accused of having rubbish memories, and their funerals often involve the indignity of being flushed down the loo. Perhaps they might garner a little more respect if it was more widely known that goldfish are actually classical music connoisseurs.

In 2013 a Tokyo study trained four fish to show their powers of recognition by giving them a nice bit of food to chow down in return

---

* To use an analogy from our friend The Blowfish, 'If you want to understand what it feels like to be a neuromast (and why wouldn't you?) go and stand on a train platform. Before the train pulls in to the platform you'll feel a wave of air hit you and you'll know that the train is coming.'

for tugging a red bead in their aquarium whenever they heard a certain piece of music. Two of the fish were repeatedly played twenty seconds of Bach's *Toccata and Fugue in D Minor*, while the other pair were played Stravinsky's *The Rite of Spring*. Each time they were played a different section of music, which meant that they never heard the exact same part of the piece twice. Once trained up, the testing began and the fish were either played a snippet of the music from the familiar piece or one they had never previously been exposed to. If they knew the tune they were to bite down on the bead; if it was unfamiliar, they shouldn't.

Incredibly, in three-quarters of the trials the fish knew the difference between their Bach and Stravinsky. Even more amazingly, further tests showed that some individual fish actually indicated a preference for one composer over another, a bit like us humans.

But goldfish and people aren't the only animals that can distinguish between the classical greats. An earlier study, back in 1984, showed that pigeons could do the exact same thing and in 1998 Java sparrows were shown to have their own distinct personal musical taste. Though the birds all preferred classical music over modern, some had their own favourite classical composer, even exhibiting this preference when brand new pieces by their fave composers were played for the first time.

Parrots also have quite varied taste in music, as was discovered thanks to three African Greys tested in a series of trials in 2012. All three loved U2, Joan Baez and UB40, dancing and singing along to them with parrot calls and human words (although disappointingly, no lighters in the air were observed). They all liked a cheeky bit of Bach, too, chilling out and preening themselves any time it was played. Two of the parrots were then given their own personal little parrot jukebox with beak-touch screen monitors and allowed to choose what tunes they wanted to listen to for a whole month. One kept plumping for the poppy 'I Don't Feel Like Dancing' by

Scissor Sisters, while the other went for classical treats like Vangelis's 'La Petite Fille de la Mer'. However, both showed a deep dislike for dance/electronic acts like The Prodigy and The Chemical Brothers, squawking out in distress whenever they were played. Parrot-curated radio is surely the next logical step.

Alongside goldfish, Java sparrows and parrots, rats, koi carp and pigeons, have also demonstrated the ability to differentiate pieces of music from each other. And that's not all: those smart little pigeons seem to be the real culture vultures of the avian world, not only able to distinguish between musical greats, but between painterly masters as well.

Could Pigeons Be Art Critics?

Some people think that pigeons spend their days hanging around Trafalgar Square plotting over who to poo on next. But could they be more interested in what's going on *inside* the National Gallery than outside?

In 1995 a pivotal study proved that pigeons could tell the difference between works by Picasso and Monet, and even recognise which of the artists were responsible for paintings presented to them that they'd never seen before.

The pigeons were divided into two groups, one for each painter, and trained to correctly peck when they saw work by their group's chosen artist. Ten consecutive correct identifications of either Picasso or Monet paintings would be rewarded with some delicious hemp seeds, but a peck for the wrong artist would get the pigeons nothing. Over time the birds learned to recognise which paintings were the key to all that hemp-seedy goodness – but here comes the best part.

Once the birds were properly trained up the researchers started to introduce brand new paintings. It's one thing to recognise the specific arrangement of brush strokes for a single composition, but could the birds recognise the signature style of their designated painter's works? Surely that would require greater computational power than a mere bird brain is capable of?! Incredibly, almost all of them did, even when elements such as colour were taken out of the equation, showing that pigeons have an impressively complex visual processing

system. Even more astounding was the fact that they were able to group together painters of a similar tradition – lumping Impressionists Monet, Cézanne and Renoir into one bracket, and Cubists Picasso, Braque and Matisse into another.

Interestingly enough, pigeons could still recognise a Picasso painting when it was upside down, but struggled if it was a topsy-turvy Monet, proving perhaps that pseudo-art lovers who accidentally hang their Picasso prints the wrong way up may have a point – Picasso *can* be enjoyed from any angle! – while less surrealistic paintings, such as those by Monet, should never be messed with. The more realistic depiction of objects in Monet's impressionist genre clearly aided a pigeon's visual recognition in ways that their more abstract counterparts did not.

GEEK CORNER: Pigeons have amazing visual systems, possibly even better than ours. They can see in five different colour bandwidths (as opposed to our three) and, unlike us, don't fill in visual information when certain parts of an image are missing. They have incredible visual memories, able to distinguish between more than 1,800 memorised images.

So far, so impressive, but the chief scientist behind the study, Professor Shigeru Watanabe, wanted to take things further. In 2001 he compared the ability of pigeons vs humans to distinguish between the works of Chagall and van Gogh. Again, half the pigeons in the study were rewarded with tasty food in return for pecking Chagall paintings and half in response to van Gogh. The pigeons were then shown three never-before-seen paintings by both artists. They successfully managed to tell which was which, even when half the painting was covered or the colours completely removed. The (human) men and women tested were no better or worse than their grey-feathered counterparts, proving that when it comes to spotting artistic style pigeons may be just as adept as us!

Clearly pigeons have the ability to recognise and remember defining features of artists' work ... but artist-spotting does not a critic make. To really give their human counterparts a run for their money they'd have to be able to tell *good* art from *bad*. So, our old friend Prof. Watanabe put pigeons to the ultimate art critic test in 2009. Schoolchildren's artwork was judged by ten adults (including their art teacher) to be *good* or *bad*, depending on how clear and recognisable the pictures were. Pigeons were trained in a similar way to previous experiments – this time pecking when paintings were *good* versus *bad* – following the judgement of the adults. Again, once trained up, the birds were then shown novel paintings by the kids. Lo and behold, they were able to pick out the ones that the human raters had deemed *better* than the others.

What does this all mean? Are pigeons really as artistically savvy as us? Can pigeons learn to recognise beauty in the same way that we can? Not quite, because as of now pigeons aren't able to indicate whether a painting with a completely *new style* is any good or not, meaning if we relied solely on pigeons as art critics we'd never appreciate any groundbreaking trailblazers who push art in new directions. How boring that would be.

That said, Watanabe's work is far from finished. Who knows what else our feathered friends are capable of? We don't know about you, but we'd like to see a flock of pigeons let loose at next year's Turner Prize – choosing the winner with tiny little pecks of their beaks.

Pigeons are clearly more intelligent creatures than many of us humans give them credit for and, perhaps more surprisingly than their art and music savvy credentials, a more recent study established whether or not they could also be put to use as pathologists . . .

Pigeon Pathologists

If you've read this chapter in sequential order like a good, conscientious little egg, then by now you'll know that pigeons are much more than just walking, squawking, deformed-footed little poo machines. However, these culture-loving birds have yet one more trick up their grey-feathered sleeves: they could even be the pathologists of the future.

'How so?' we hear you say, as you fumble for your Valium. Well, scientists managed to teach a flock of pigeons to spot the signs of breast cancer by training them up on images of real patient biopsies. The sixteen birds were shown various slides of breast tissue biopsies and had the option of pecking a yellow or a blue button. Press the correct button when signs of malignant cancer were on the screens and they'd get a tasty treat. Get it wrong and they got nothing. After only a few hours the birds were already starting to correctly identify cancerous cells and after a month their accuracy was up to 80 per cent. That's not quite as good as their human counterparts, but when combining the entire flock's responses their success rate flew up ('scuse the pun) to 99 per cent. Incredibly, that's on a par with human pathologists and *much* better than current computer image analysis performance.

Does this mean that pigeons are an attractive proposition as an inexpensive supply of pathologists? After all, their training is short and food-based-salaries much cheeper (and again) than ours. The chief researcher in the study reckons probably not. Job-stealing super-computers are far more likely to do both the human pathologists and the pigeon newcomers out of the job. Back to pooing on statues in Trafalgar Square for our pigeon buddies after all, then.

Pigeons aren't the only ones that can be taught to spot cancer — man's best friend has also be trained to use their hooters to sniff out early stages of the disease.

Dogs That Sniff Out Cancer

Ever suffered the embarrassment of a dog sniffing away at your bum? Kind of annoying, but if you knew that our canine buddies were actually capable of smelling out cancer, might you be willing to forgive them for the occasional olfactory indiscretion?

The idea that the presence of disease could have a distinct odour is not a new one. Around 400 BC Hippocrates stated that doctors' 'nostrils indicate much and well in fever patients, the odours however differ a lot', and Chinese medical texts dating back to the third century BC had it that 'every disease of the five solid organs is reflected in colour and smell'. If true, it makes sense that dogs would be the ones to sniff it out, famous as they are for their incredible sense of smell. This olfactory over-performance is thanks to the thirty million scent receptors in the snouts of most dog breeds compared to the measly six million we have in our humble human noses.

GEEK CORNER: In 2015 scientists proved that an Australian woman named Joy Milne could smell Parkinson's disease (PD). She first smelt a change in her husband, who was later diagnosed with the disease, and was then put to the test in the lab. Twelve subjects (six with PD and six without) wore T-shirts for the day, which Joy would then sniff. She correctly identified eleven of the twelve, insisting that a 'control' subject also smelt of PD. Incredibly, but also tragically, the control subject was diagnosed with PD eight months later, proving that Joy was 100 per cent accurate in the original testing.

In 1989 a British woman, who later became the subject of a BBC documentary, made an observation that seemed to back this idea up in a truly extraordinary fashion. She told her doctors that her dog kept barking and nipping at a dodgy-looking mole on her leg, which later turned out to be cancerous – a phenomenon which has since been reported by many others pet owners. Such anecdotal stories were enough to inspire scientists to really put it to the test, so they began putting the Labs into the labs, so to speak.

The first study in 2004 tested ordinary dogs of six breed varieties, training them to sniff urine samples of patients. They managed to detect bladder cancer in 41 per cent of cases, far higher than the 14 per cent hit rate you'd get through blind chance alone. But some dogs seemed to have better schnozzles than others, so later research put the best breeds to the test once again. This time they found that these superior dogs could successfully sniff out prostate, breast, colorectal and lung cancer in samples with up to 99 per cent accuracy.

CHIC FACT: Dogs are now being trained to sniff out bee diseases to help bee-keepers protect their hives (see pp.189–92).

So what exactly are the dogs' hooters picking up on? Scientists reckon it may well be 'volatile organic compounds' that they're detecting, produced by malignant cancer cells as tumours grow and are eventually excreted into the outside world. Dogs seem to have an uncanny ability to sniff out these volatiles, even when the cancers are in the early stages – the time when detection is most crucial to enhance chances of survival.

While further testing is needed, many believe that doggy disease sniffing could very well be a way to complement traditional methods of diagnosis in the future. With dogs already saving lives daily by sniffing out bombs, drugs, weapons and people buried alive in earthquakes or avalanches, this new cancer scent-detection string to their bows proves once again that dogs really are our best friends.

Clearly dogs do us humans a lot of favours, but when it comes to actually loving us who does it best – dogs or cats? Let battle commence.

Who Loves Us More? Dogs or Cats?

It's the eternal question that divides nations – are you a cat person or a dog person? Well, cat lovers be warned – the next study may have you protesting in droves.

Scientist Dr Paul Zak carried out the research in question for a TV documentary in early 2016, measuring levels of oxytocin (the so-called 'love drug') in pets' bloodstreams after they had spent some time playing with their owners. In humans, oxytocin is released by the hypothalamus when in the presence of our loved ones and rises significantly when we are kissing, touching or breastfeeding. We also get a hit when stroking our pets or looking into their eyes. Earlier studies proved that dogs feel the same, but Dr Zak wanted to know how our feline friends compared?

Twenty pets (ten cats and ten dogs) were put to the test, with saliva samples taken and oxytocin levels measured before and after ten minutes of playtime with their owners. They found that, while dogs' levels rose by nearly 60 per cent, cats shifted up a measly 12 per cent. This suggested that dogs had a far stronger physiological response to their owners' attention than cats, which many would claim proves once and for all that pooches love us a whopping five times more than our feline friends. Damn you, Tiddles!

CHIC FACTS: Goats are the surprising, and so far undisputed, biggest dog lovers of all. When Dr Zak let a dog and goat spend time together the goat's oxytocin levels shot up by 210 per cent.

However, all is not lost for the cat lovers among us, as raised levels of any amount at least disproves the doubters who claim that cats have no true affection for humans at all. Further studies also showed that cats feel less anxious than dogs when separated from their owners, again suggesting that dogs are more attached to us than their feline counterparts. But, while cats may, as Rudyard Kipling once suggested, walk by themselves, we humans can still gain huge benefits from a dose of kitty love in our lives. They can make us less stressed and less lonely for starters, and new research has shown that even just

watching videos of cats may be an effective way of lightening our mood. Tiddles, we take it back. All is forgiven.

Hoorah for dogs, who can sniff out all manner of potentially human-harming substances, diseases and even seem to love us more than cats. But, every silver cloud has a murky grey lining . . .

Jealous Pooches

Loving someone can be wonderful, but it can also have a bitter side. If you've ever let jealousy get the better of you, you'll sympathise with our pooch friends here, because dogs can fall victim to the green-eyed monster, too.

If we're being completely honest, all of us have experienced envy in various forms throughout our lives, whether it be over a sibling, a friend or even an inanimate object, but can dogs really show this same level of complex psychology? After all, the ability to measure and compare how other animals feel towards a special someone and vice versa — which is clearly a prerequisite for jealous behaviour — involves an impressive array of cognitive skills for monitoring and comparing social relationships between others. Could dogs really comprehend the potential threat to their relationship when a third party is introduced into their dog–master relationship? According to one American study, they certainly can.

CHIC FACT: Beware! Jealousy is a dangerous thing. It's thought to be responsible for a third of all murders across the globe.

Thirty-six dogs were observed as their owners took part in various activities. The dogs were pretty chilled when their masters ignored them or took part in other activities such as reading aloud from a book. But when a rival in the form of a stuffed dog toy was introduced into the mix it was a whole different story. A quarter of the dogs became agitated, growling and snapping at the faux-dog as it wagged its false tail and barked for effect whenever petted by the owner. Some would also try to get in between their master and the toy, attempting to regain their beloved owner's attention.

The study largely mimicked an earlier one by the same researcher showing that six-month-old babies got similarly jealous if their mums played with dolls, suggesting that it is a primal emotion that may have evolved to protect those relationships essential to our survival.

CHIC FACT: Anecdotal studies showed that 81 per cent of dogs and 79 per cent of horses showed signs of jealously when other animals were in the presence of their owners.

What does it all mean? Are dogs really capable of these complex emotions? Only time, and more research, will tell, but for now be aware: when Lassie's in the room don't go giving newbie Rover all your attention.

So dogs can get jealous, but can they also hold grudges? Dogs are starting to sound more and more like humans by the second . . .

The Dogs That Hold Grudges

Have you ever sat in a room and felt that your friend's dog was giving you a bit of a dodgy vibe? Maybe it's their general demeanour, their lack of affection or a reluctance to take a treat from your hand. You feel as if you're being judged, but surely you're just being paranoid?

Well, ask yourself this: have you done anything in front of the dog it might have perceived as being offensive to its owner? Maybe you refused to hand over the remote control for the telly or help them out with a simple task. If so, it's very possible that their dog may in fact be holding a grudge against you.

Sounds crazy, we know, but according to a Japanese study in 2015 it could be true. The researchers set up a room where dogs could watch their owners attempting to open a box in the presence of two strangers. For the sake of argument (and our own personal pleasure) let's call our example dog Snoopy, the owner Charlie Brown and the two strangers Peppermint Patty and Woodstock. Charlie Brown would attempt to open a box, while Snoopy looked on and Woodstock got on with his own thing in the corner. Charlie would ask Peppermint Patty to help him with the box, and she'd either rudely refuse or get stuck in.

So far, so good. But next up in the experiment Charlie Brown would leave the room, at which point Woodstock and Peppermint Patty would both offer Snoopy a dog treat to see who he'd prefer to take the treat from. The result? If Patty had refused to lend Charlie a helping hand Snoopy would snub Patty's treat and head straight over to Woodstock.

OK, so in the real study not *every* single dog was called Snoopy, and not every single dog chose the neutral stranger's treat over the one who'd been unhelpful to the owner. But over the course of the whole experiment the results came out with a statistically significant difference: overall, dogs definitely preferred to chow down on treats from someone who they *hadn't* seen rebuffing their owners' pleas for help. On the other hand, when the stranger *did* help their owner with the task in hand, the dogs were equally happy to take a treat from either of them.

> **CHIC FACT:** Aside from dogs and humans, a similar study to the one above found that some primates such as capuchins show loyalty, without there being any obvious direct benefit to them. Surprisingly, our closest relative, the chimp, doesn't show this trait. Selfish buggers.

Could it be, as researchers believe, that dogs aren't purely self-interested (even when it comes to treats) and that, more importantly, they can make 'social and emotional evaluations of people'. Human beings can start to do this by the age of about three, but until now it was thought to be way beyond canine abilities – yet more proof of their ability to build a highly social and collaborative society.

It turns out we have more in common with dogs that we thought, so the next time you consider being rude to a mate in front of their four-legged friend – maybe wait 'til Snoopy has left the building or you can expect a serious canine-snubbing down the line.

Though dogs will selflessly snub treats from rude strangers in the name of loyalty, other species are less noble. Many will go to great lengths to ensure they get their dinner, even if it means screwing over their fellow creature. The most nifty of these food-nicking techniques we've come across was from the dark lords of the night sky themselves – bats.

Holy Bat Jamming!

Batman hasn't got it easy – saving the lives of the ordinary folk trying to go about their day in the depraved city of Gotham takes a heavy toll. And his man on the inside, police Commissioner Gordon, has it pretty rough as well. Gotham is one dirty, dirty city but between the two of them they have it pretty much nailed when it comes to keeping the bad guys at bay.

You see, although Commissioner Gordon and Batman have had a fractious relationship at times – the Commissioner reluctant to have to rely on a vigilante hero, the hero himself feeling undervalued – they have always put their trust in a vital piece of equipment they just can't do without: the Bat Signal. This giant bat-shaped searchlight calls Batman to action in the very worst of scenarios, be it the streets

being flooded, the city bank bursting into flames, or all the animals being let loose from the zoo – it's the way that Commissioner Gordon puts out a message asking Batman to get his spandex on and save the city, once again.

So imagine for just one moment that their arch-enemy, the Joker, cottoned on to what could potentially be Batman's greatest nemesis, a nemesis from which the hero takes his own name – an actual bat! What if, when the Commissioner flicked the switch on and when the light started to emanate towards the sky it stopped halfway to the clouds, a tiny black figure hovering over, blocking the beam? What if one day the Joker realised that he could use a real bat to jam the bat signal?! Batman wouldn't get message that the villains were wreaking havoc in the city, our vigilante hero wouldn't be able to 'wham!', 'bop!' or 'kapow!' his way to glory, and Commissioner Gordon would certainly get the sack. Holy Jobseekers!

Now we don't want to give any evil geniuses any new ideas, but life is often stranger than fiction, and believe it or not here in the real world bats jamming each other's signals is common place (and just as dastardly as any evil plot The Joker could come up with). Here's how it works: firstly, if you're a bat then, given that you're blind and everything, you're really not interested in beams of light creating bat-shaped silhouettes on the underside of clouds. Sound, not light, is your modus operandus: a brain that evolved over many years to emit a high-pitched squeak and use the various echoes bouncing back to navigate, communicate and hunt for tasty insectivorous prey, sonar-stylee.

The thing is, you're not the only bat in the village and with populations in some residences numbering in the millions, competition for flying insect snackage can be fierce. In these kinds of situations you need to dream up an ingenious way to make your batty competitors miss their opportunities to pounce on their prey, to give you half a chance to strike it lucky yourself. Long and short of it: you need to screw over your competitors to make your own survival more likely. Altruism isn't really big in the bat community.

The Mexican free-tailed bat is just one example that finds itself in these kinds of competitive, overcrowded circumstances, but how exactly do they (you're off the hook now, and back to being a human) achieve their Machiavellian mission?

Do you remember the kid at school who used to wait until the bell went and then come and steal your lunch money? Well, bats employ a not-too-dissimilar technique in many ways.

CHIC FACT: Bats get a bit of a bad rap, blamed for everything from rabies to blood sucking and inspiring tales of Dracula with their tiny pointed teeth. In reality, fewer than ten people have contracted rabies from bats over the past half-century and the majority of bats actually prefer to eat insects. However, there are three species of actual vampire bats. They do like to drink blood, Dracula-style, but prefer to lap it up, and will bite a cow, or even *you*, to get to the good stuff.

The lunch bell in our *new* analogy is actually a series of chirps that hunting bats emit as they close in on their prey. The chirps get steadily faster the closer they get, and the feedback from the noise bouncing back off the insect helps the hungry bat work out exactly where dinner is in relation to them so they can grab the tasty morsel out of thin air. Unfortunately, this distinctive sound, known as the 'feeding buzz', also alerts any other greedy nearby bats to the fact that they are zoning in on something tasty. The sneaky rival now has the perfect and tantalising – if not ever-so-slightly mean – opportunity to scupper the original hunter's chances.

GEEK CORNER: Bats can be handy in countries where mosquitos are a problem as they've got killer appetites – one bat can eat a whopping 1,200 insects an hour. In the Bracken Cave in Texas live around twenty million Mexican free-tailed bats, which collectively get through around 200 tons of insects every single night.

The rival bat emits a similarly high-pitched sound that confuses our original hungry bat (effectively by jamming its echolocation circuits), so that hungry bat can no longer pinpoint the prey's location and misses its target. The rival bat nabs the prey instead, leaving the original hungry bat to go hungry.

This theory was put forward by a team of American scientists (although admittedly they didn't use the terms 'hungry bat' or 'rival bat') who were studying wild bats in Arizona and New Mexico and decided to follow up on an earlier study that had proved that tiger moths, in order to escape being munched down, make high-pitched sounds to confuse bats in a similar manner. In fact, the moths' jamming signals were so effective that some bats missed even when the moths were tied down!

Hot on their heels, the new team spotted that bats only emit certain characteristic calls when another bat was making its 'feeding buzz', aka the treacherous dinner bell. Realising that this could have simply been one bat saying to another, 'Oi – hands off! This is my territory and that's my grub, scumbag!', the team decided to investigate.

In scenes reminiscent of many a cruel little boy's playtime the researchers tied insects to a long piece of microwire between two trees in bat-hunting territory. They soon noticed that hunting bats would miss their prey far more often (by 70 per cent) when what turned out to be the bats' 'sonar jamming' calls were blasted at them from speakers during their 'feeding buzz'.

CHIC FACT: Bats aren't the only animals willing to resort to underhand tricks to get some extra food – capuchins are masters of food sneakery, mimicking wild cat sounds to scare the elders away to grab the grub for themselves. Drongos are also devious little blighters. The birds act as security guards for meerkats, giving the alarm when predators are nearby. After a little while, if the crafty birds spot a particular tasty treat, they sound a false alarm, scaring off the meerkats so that they can swoop in to nab it for themselves.

It's a shame that bats never quite figured out how to apply this principle to humans. It would have made for entertaining viewing had they managed to emit a sound that made Ozzy Osbourne miss the poor infamous bat's head and chomp down instead on his own hand live on stage.

Bats may be skilled food thieves, but they haven't yet mastered the art of sneaking themselves a free round. Some creatures, such as chimps, however, have figured out a way to steal booze from their neighbours, even when those neighbours happen to be smart creatures like us.

Drunken Monkey

Remember the classic scene from Disney's *The Jungle Book* where Mowgli is kidnapped by a group of mischievous orangutans who whisk him off to see King Louie in the temple, who promptly dances about and sings:

'I wanna walk like you, talk like you, get drunk like you-oo-oooo'

OK, OK – those may not be the *exact* lyrics, but perhaps they will be in the future, because in June 2015 a group of British and Portuguese scientists confirmed what had been suspected for many years, that – just like us on a Friday night – some of our primate cousins have a penchant for getting boozed up.

CHIC FACT: The first official observation of a 'drunken monkey' was made in 1779 by a doctor aboard the HMS *Dorchester*. He wrote extensively in his diaries about monkeys 'cavorting madly' on a beach 'in a state of drunken delirium'. The sailors with them even considered trying to find and steal the monkeys' booze supply, but when they tried the monkeys fought for their stash and 'launched a furious assault'. Sadly, five monkeys were killed and two men were mauled, proving, yet again, that alcohol can make idiots of us all.

The study itself had been going on for quite some time (1995–2012), with scientists observing West African chimpanzees engaging in a tipple on a surprisingly frequent basis in the forest of Bossou (located in the south-eastern corner, Guinea, right next to the borders with both Liberia and Ivory Coast).

GEEK CORNER: The fermented sap from the study in Bossou was no shandy: the average ethanol content of the sap was 3.1 per cent ABV (alcohol by volume), or the same as a pint of a Greene King dark ale. The strongest sap brew reached as high as 6.9 per cent – which is 1.4 per cent more than Moscato wine.

While the wild chimps were being filmed in their natural habitat they were caught sneakily dipping into supplies of alcoholic sugary

palm sap, which was being harvested by local villagers to make palm wine. Locals would tap the raffia palm trees at the crown to collect the sweet sap – which quickly ferments into a boozy brew – in containers below, covering them up with large leaves to keep away pesky little bugs, and leaving them to fill up. Unfortunately for the villagers, they didn't predict that a rather larger and smarter creature had worked out how to get into their supply.

During fifty-one separate recorded incidents wild chimps were observed helping themselves to the tipple using rudimentary tools they'd fashion for just this purpose: they would chew large leaves to create either leaf scoops or, more commonly, leaf sponges, then dip these leaves repeatedly into the containers and suck the booze off them. Some of the chimps would even indulge in some serious drinking sessions, with a few kicking off as early as seven in the morning. And we thought Keith Richards was rock 'n' roll.

This isn't even the first time our hairy cousins have been caught getting smashed: the slow loris is also partial to indulging in a cheeky bit of fermented nectar (its' tipple of choice is from the bertam palm) and vervet monkeys have been filmed polishing off cocktails on beaches in St Kitts before staggering around, off their faces, like a bunch of British tourists on their summer holidays in Magaluf.

Why do our cousins engage in such debauched behaviour? The 'drunken monkey hypothesis', first mooted in 2004, suggests that primates that are attracted to ethanol – booze to you and me – were once more likely to survive and reproduce, giving them an evolutionary advantage. The reason? Fermenting fruit was once a primary food source, and alcohol stimulates our appetite and encourages us to seek out more food, something anyone in the Friday night kebab queue can no doubt relate to.

If you thought that the strange antics of animals on our lands and in our skies were bizarre, then wait 'til you hear what our fishy friends have been getting up to. From farting herrings to the harlequin shrimp butchers and the gonad-munching anus fish, marine life may well be the freakiest of them all.

Under the Sea

The sea is the most mysterious part of our planet. Despite covering more than 70 per cent of Earth, we have only managed to explore a measly 5 per cent of it, meaning many of its secrets still remain locked up or hidden somewhere in its watery depths, waiting to be discovered one day.

Exploring the deep blue sea is a task riven with problems – it's time-consuming, expensive and often dangerous. Yet every year curious marine biologists (and seafaring robots) reveal some extraordinary, strange, eccentric new creatures from the deep, many of which blow our tiny little minds. Whether it be the ghostly, spaghetti-legged deep-sea bigfin squid – that looks like an alien with limbs up to 26 feet in length – the stunning luminous-orange Venus fly trap anemone, or the ancient-looking Lazarus fish – so-called because it 'rose again' after being presumed to have become extinct sixty-five million years ago – marine discoveries can be truly mind boggling. They can, however, also be ridiculous, funny and downright implausible. Here are some of our personal faves.

Farting Fish

You may not be keen to admit it but we all fart. Everyone does. Yup, even your squeaky clean Geek Chic author duo. It's a perfectly natural and vital bodily function, albeit a pretty smelly one.

Deep down under the sea is probably the last place you'd think farting was going on. Sure, we might have let a few cheeky ones rip underwater and then tried to blame it on a few stray air bubbles from a passing shoal of fish, but did you know we aren't the only ones who suffer from occasional excess flatulence? Our scaly, gilled little buddies can also let off little Tommy Squeakers into the deep blue sea.

Fish, you see, have swim bladders – like little balloons inside them that can either be filled with air or emptied – to control their buoyancy. If you've ever been scuba diving you'll know what we mean: every scuba jacket comes with its own air control buoyancy regulator to help divers submerge deeper, or head back towards the surface. So fish have the same things; theirs just happen to be inside them. And there are two types – physostomous and physoclistous. Physostomous bladders are the ones we're interested in, because they can open up and take in air from the gut. When a fish wants to have a little floating aid it heads to the surface and take a few gulps of fresh air, which fills up its physostomous swim bladder. When the same little fishy later wants to get rid of air it forces it back into the gut and out of its bottom. And there you have it: living, breathing, farting fish.

And that's not the half of it. Some fish, such as herring, use farting as a form of communication – not unlike a few drunken students we knew in our uni days. The sound that comes out of them is like a squeaky raspberry. The gases emitted are also coated in oils from the gut, acting as pheromones to transmit chemical messages to nearby fish.

The noise and the smell of the fishy farts could serve as a warning to other fish of nearby predators, a means by which fish can find each other and stick together in the dark, or might even be used as a means of calling out to potential mates.

It's certainly a novel way to woo a new lover, farting in their direction and hoping it reels them in, but then again we imagine fish would probably think swiping right on Tindr is quite a strange way of finding a potential mate, too.

Anus Fish

With ever-rising rents and a looming housing crisis, many of us today are compelled to look for alternatives to traditional accommodation. Houseboats, squats and mooching on friends' couches are a few of the options available to us humans, but spare a thought for the poor old pearl fish, which often finds its home in the crevices of another creature's anus. Dark times. Dark times indeed.

The lucky creature in question that gets its anus inhabited by the tiny parasitic fish is the sea cucumber, which looks a bit like a giant bloated caterpillar with a mouth on one end and a bottom on the other. They're very basic beasties, spending their lives slowly cruising

along the bottom of the seabed, hoovering up sand and edible matter before pooping it out the other end. In fact – delicious fact coming up here – the beautiful white sandy Caribbean beaches you may have holidayed on? It comes out of various sea creatures' bottoms, mostly the coral munching parrotfish. So, in a way, all those pristine white, sandy beaches are made of fish poo.

Back to our pearl fish, because anus bothering isn't the only thing our poor simpleton sea cucumber friends have to contend with. These nasty little fish slowly eat them from the inside out, starting with their gonads, before moving onto the rest of them.

Ready for your lunch yet? Cucumber sandwich anyone?

The Harlequin Shrimp Butchers

Thought Jack the Ripper was terrifying? You ain't heard nothing yet. In terms of cruelty, the harlequin shrimp butcher really takes the sadistic biscuit.

As with many savage beasts, harlequin shrimp are also stunning creatures. They pair up with their lover and then go on the hunt for their prey, like a shrimpy Woody Harrelson and Juliette Lewis in *Natural Born Killers*.

One they've located their victim – the blue starfish – the shrimp pair get to work. The killer team first pull the starfish from a rock and then flip it bottom side up, rendering it helpless. Next, they haul it onto their backs and drag it back to their burrow, where they slowly proceed to dissect it. What's more, they keep the starfish alive throughout the course of days or even months during which time they gradually nibble at it piece by piece.

The worst part of all is that starfish, if properly fed, will regenerate and grow back missing limbs. So there's no reason why, once the shrimp butchers have the starfish back in their evil lair, they can't keep it there as a handy, living, ready-made larder, providing fresh food for ever, and ever and *ever*.

And we thought *The Texas Chainsaw Massacre* was the stuff of nightmares.

CHIC FACT: All shrimps are born male, becoming female as they mature – which is a little reminiscent of the old view that women were just underdeveloped men (see p.161)

We've covered creatures of the land, the air and the sea, but the last curious animal we'll be discussing before we move away from the animal kingdom is the earworm. And if you feel that we had to use a rather large literary shoehorn to fit the science behind the torture of having music loops stuck in your head into a chapter about animals – you'd be spot on! But, screw it, it's our book and we'll digress if we want to.

Can Chewing Gum Cure Earworms?

It's an all-too-common occurrence – you're walking down the street, quietly minding your own business, and then, out of nowhere, 'it' pops back into your head. That same old song you've had whirling round and around your brain for days; round and round and round AND ROUND it merrily goes, driving you to distraction in the process. No matter what you do, you just can't seem to dislodge it from your mind's ear.

Welcome, friends, to the dubious joys of the 'earworm'.

We all get them – Lliana tends to get a bit of Kylie Minogue's 'I Should Be So Lucky', while Dr Jack gets Dick Van Dyke singing 'Chim Chiminey, Chim Chiminey, Chim Chim Cherooo'. Apologies if we've passed those ones onto you now.

CHIC FACT: The word 'earworm' comes from the German *ohr* and *wurm*, which translates as, you guessed it, 'ear' and 'worm'. This arose as a result of a historical quirk of medicine: in ancient times, dried and ground up animals were commonly used to treat ear diseases!

And like so many of you no doubt reading this now, we couldn't work out how to get rid of the most pesky of pests.

That is, until 2015, when researchers at Reading University discovered that the best way to block these repetitive involuntary musical memories is actually ridiculously straightforward. The solution is even simpler than many of the traditional remedies, such as reading a book, singing another song or doing an attention-absorbing brain-teaser. All you have to do to purge an earworm, it seems, is chew gum.

The reason is that chewing gum uses the same muscles you engage during sub-vocalisation – saying words out loud in your head, for example, when you're planning a speech or thinking about what you might say to your boss when you pull your latest sickie. By moving your jaw it is thought that you automatically initiate the sub-vocal rehearsal mechanism in the brain, which gazumps the earworm. Why? The most likely explanation is that, rather than generating the sound of the offending tune, your auditory cortex starts preparing the sound of the words you might be about to say

instead. Chewing gum on its own is enough to trip that circuit, even without any actual sub-vocalisation – *et voilà*, earworm successfully jettisoned!

So next time you hear someone complaining that they're being tortured by a catchy little ditty that's stuck in their head and won't go away, just reassure them that help is at hand and pass them a stick of Wrigley's finest.

> **CHIC FACT:** A bundle of earworm facts: earworms are also referred to as 'brain worms', 'stuck song syndrome', or even – as the leading academic behind this particular study calls them – 'tune wedgies'. 98 per cent of people experience earworms. Women tend to experience them for longer durations than men. Most earworms are loops of between fifteen and thirty seconds and tend to be songs with lyrics and simple melodies. Murderer Jean Harris was said to have the same earworm for thirty-three years, regularly hearing 'Put The Blame On Mame' in her head after first hearing it in the film *Gilda*.

Final Thoughts

There you have it – not so much all creatures great and small, as all creatures great and drunk and cruel and clever and jealous and musical and loving and . . . flatulent. Our fellow animals certainly are a funny old bunch. In many ways they act (almost) as bizarrely, inappropriately and ridiculously as us.

5

Paranoid Android

The robots are coming . . .

Not in the sense that they are going to chase you down the street or anything. More that, after a few decades of rapidly accelerating technological innovation, robots will soon become a very normal part of everyday life.

In fact, everyday robots have already infiltrated homes across the globe. The world's largest robot manufacturing company, iRobot, has sold over fourteen million household robots alone. Their most accessible product is the dust-munching vacuum cleaning Roomba robot, which uses specially designed sensors to avoid bumping into walls, pets and furniture as it zigzags across the floor.

Outside our homes robots have long been key to various manufacturing processes, like building cars. Underwater robots clean pools for the domestic market and mend pipelines deep under the sea for corporate entities. Up in our skies, drones buzz around filming footage and dropping bombs, while scientists send them off to remotely investigate harsh environments all over the world – and beyond.

Meanwhile, androids (human-like robots) are becoming so visually realistic and their conversational capacities so fluent that they're being touted by experts as the most likely candidates to look after us when we're old and grey and may *even* end up being our lovers and spouses!

More worrying still, in 2015 more than a thousand robotics experts got together to sign an open letter outlining the need to limit development of autonomous-weaponised drones. At the moment, when a drone aims at a target it is a human that ultimately makes the call on whether or not to pull the trigger. However, machine learning is rapidly approaching the point where a drone could be armed with software sophisticated enough to enable them to automatically recognise the tell-tale combination of features that accurately identify a person's ethnicity. There are now serious concerns about the possibility of ethnic cleansing on a frightening scale should such technology fall into the wrong hands. It's hard to think of anything more horrifying.

Sci-fi writers have been warning us about the threat this could ultimately pose to mankind for decades. Films like *Robocop*, *I, Robot* and even *Wall-E* have gently nudged us into an awareness that the age of the robot may soon be upon us and, if we're not very careful, it could all end in tears – and blood, for that matter. Modern-day robots may not be quite up to the level of sophistication and menace imagined by Hollywood scriptwriters, but they are certainly starting to head in that direction.

Should we fear for our lives? Or are robots just the innocent, metallic answers to all our labour-saving prayers? The answer may be a combination of both, but make no mistake: robots will soon become an intrinsic part of our day-to-day lives. One thing's for sure, the robots of tomorrow aren't going to look much like Metal Mickey, R2D2 or that grumpy one from *Lost in Space*. They'll sneak into our lives, in disguise, as easily as an automatic can opener or robotic vending machines in our places of work, leisure or study. They'll pop up out of nowhere to provide us with quick, effective, seemingly effortless service that, soon enough, we'll feel we cannot live without.

Over the course of this chapter we'll peek behind the Wizard of Oz's curtain and share with you some of the key features of today's most incredible robots. We'll look at some of the most mind-blowing and thought-provoking new developments in modern-day robotics, take a sneak peek at what the future of robots has in store for us *and* we'll merge all this geekery and chicery into a vision of our ideal imaginary robot.

Robot Swarms

There's a real buzz in the robotics community at the moment and it's coming from swarms of mini-robots, which cooperate a bit like insect colonies, combining forces to achieve incredible things.

At first glance the term 'swarm of robots' doesn't seem tremendously inviting, we'll be the first to admit, but rest assured they are very impressive. One mob of roboticists have called their inventions the 'Kilobots', winning them our prize for 'Best-Named-Robot-Swarm-To-Date' (it almost doesn't matter what they actually do given that, should people overhear us chatting about them in the pub, they'll think we're saying 'Killer Bots' and start freaking out that the Terminator's coming). The choice of name actually reflects the sheer numbers of individual automatons involved. In reality they're actually quite possibly the least-threatening looking things you've ever see. This population of autonomously assembling mini-robots – brainchildren of the geekers and chicers at Harvard's awesomely named Wyss Institute for Biologically Inspired Engineering – is pretty special. They can move, interact and coordinate themselves to form any particular shape you might desire, all at the touch of a button.*

Make a Starfish!
At the command 'make a starfish' the 1,024[†] tiny robots, scuttling around on a tripod of pin legs and measuring about two centimetres

---

* The scientists took inspiration for the kilobots from ants and social amoebas that often band together to form rafts to get over obstacles they simply couldn't overcome alone.

† The observant among you may have noticed that 1,024 is the number of bits in a kilobit (kB).

across, assemble by chatting with each other via infrared: 'Hello friend', *receiving*, 'Am I in the right place?', *nope – jog on, buddy*, 'OK, cheerio!' They have this conversation – kind of – with every kilobot they come across until they find a spot that does need occupying in order to create the desired shape, which in this case is a beautiful starfish. A brand new dynamic form of art? Perhaps, but we reckon having your very own meta-organism that can assemble itself into any imaginable shape could be pretty handy – and fun to boot. Bossily shouting 'build me a castle!', 'make me a dragon-shaped carriage!' and 'make me a mini-space rocket!' like Veruca Salt jacked up on Red Bull, would be hours of entertainment and make us feel like some sort of all-powerful magician to boot.

CHIC FACT: One lab at the University of Graz in Austria is trying to recreate shoaling behaviour in twenty fish-like robots that are all called 'Jeff'.

Plait Me a Bridge!

Swarms of autonomous flying drones have managed to achieve something that is not only pretty but also potentially very useful: weaving a rope bridge between scaffolding towers on opposite sides of a warehouse. Pretty sweet. To be fair, there were only three drones involved in this particular assembly, but they did do it completely on their own. And we reckon that so long as there are more than two robots you're perfectly within your rights to say 'swarm' without being branded a porky* peddler. You can watch this robot trio in action on YouTube.

Accelerated by the technical wonder that is time-lapse photography, it is absolutely mesmerising. Right at the end, the entire research team of super-geeks cross the bridge from one end to the other, one by one. It's not quite at the level of the Twin Towers tightrope-walker Philippe Petit (whose incredible antics you can see in the docu-film *Man on Wire*), but it's still a great test of faith in their work. Nothing unravels, and none of them tumble arse over tit to the ground, testimony to its rigorous construction. Just as well, given that they were inspired to

---

* Pork pie is Cockney rhyming slang for 'lie'. Pork pie ... lie. Get it?

invest so many years of collective man hours in this project with the specific aim of seeing it used one day in real-life emergency scenarios.

> **GEEK CORNER:** Swarm robotics is a fantastic example of biomimetics, or, as we like to think of it, copying design principles from nature. The process of evolution has had millions of years to perfect systems of coordination between different individual organisms in a colony so that they can achieve more together than is possible alone. Flocking birds can navigate more safely and using less energy in a large group than individually. Each individual fish in a shoal is much less likely to be preyed upon when sticking close to many others. Swarming insects can be so overwhelming in number that they can even scare their predators off. By studying swarm behaviours that have naturally evolved to achieve certain goals, robotic engineering can stand on the shoulder of biological giants when trying to achieve similar goals.

Flood Dams!

Speaking of disasters, how about some flood protection? The trouble with rivers is that they hold up, not leaking water year after year, lulling everyone into a false sense of security, only for an unexpected deluge of rain to come along, bursting the river's banks without any warning and catching everyone with their pants down. Water levels rise and rise. Everyone scrambles to move sandbags into place in an effort to save homes and businesses from the rising floodwater, only to find themselves too slow or too

late. Cooperative robot swarms could work together to build up walls of sandbags to act as a barrier much faster than humans ever could, so plans are afoot to develop such systems for real. Several British towns where floods are sadly becoming more and more common will no doubt be very grateful when this innovation finally becomes available.

PERFECT ROBOT UPDATE: If we're honest, we're not completely sure how we're going to incorporate the power of the meta-organism into our dream robot but we might have these mini-robots studding the surface of our big bot – ready for deployment at the touch of a button – maybe giving it a punk-style mini-bot Mohican to boot. Meanwhile, other tiny robots are busy making their mark saving lives in a different type of emergency . . .

Robo-Roach to the Rescue

Imagine this. Your worst nightmare has come true. An earthquake has just reduced the building you're in to rubble. You're trapped, alone and losing hope. Then out of the corner of your eye you see something scuttling towards you – your saviour, your hero, your knight in shining armour: a cockroach.

We know, we know. It sounds ridiculous, but this may well be a (slightly creepy) reality in the not-too-distant future, as scientists develop cyborg-cockroaches – part robot, part insect – in an attempt to speed up search-and-rescue missions in a variety of dangerous and inaccessible disaster scenarios.

CHIC FACT: Cockroaches can survive a whole week without their head. If only Marie Antoinette had been a roach, things mightn't have ended quite so badly for her when the revolutionaries of eighteenth-century France decided her time had come.

They chose roaches because they're fast, hardy and can squeeze through tiny nooks and crannies that fully man-made robots would struggle to negotiate. Hotwiring each cockroach with a simple circuit board fitted like a backpack and plugged directly into their nervous systems will enable rescue teams to control these little critters' movements and direct them into disaster zones, a bit like tiny living remote-controlled cars.

Add a tiny microphone into the mix to detect the sounds of breathing or calls of distress and rescue teams could potentially locate survivors in a fraction of the usual time. With trials already taking place in disaster mock-ups, we can expect these tiny little life-saving biobots* to save lives in real disasters sometime soon.

> **GEEK CORNER:** After roaches were rumoured to have survived the atomic bombs dropped on Hiroshima and Nagasaki in great numbers, scientists put the theory that they can survive the fallout of nuclear warfare to the test. American TV series *Mythbusters* subjected cockroaches to three levels of radioactivity and found that they survived, to a point. All lived through the lowest levels, 10 per cent survived at levels of radiation similar to those in Hiroshima (10,000 rad), but at the highest levels (100,000 rad) even the mighty cockroach bit the dust.

The next time you see a cockroach in your bath and consider squashing it, remember – they may not resemble Clark Kent in any way, shape or form, but those icky little nuclear-disaster-surviving bugs might just end up saving your life one day.

PERFECT ROBOT UPDATE: Clearly, our ideal robot will need a backpack full of the little robo-roach blighters, enabling us to access those hard-to-reach places as and when the life-saving need arises. And, better still, new developments using brain–computer interfaces mean that some cyborg cockroaches can now even be driven by thought alone. Not exactly on your bucket list? How about robots with wings?

---

* A biobot is a half-robot, half-living creature.

Robot Birds

Have you ever closed your eyes, drifted off to sleep and dreamed you could fly like a bird? It's a fantasy mankind has long harboured and one that birds achieve daily just by flapping their wings. To date, replicating this feat in machines has proved elusive – it's an evolutionary triumph that took biology millions of years to perfect, after all – but various scientists have now set their sights on creating robots that fly *just like a bird*.

Of course, many robots can already fly. Remote-controlled drones are hardly a novelty. But a breakthrough in machines flying like birds came in 2011 when a German bionics team created SmartBird. Able to fly autonomously using the power of its flapping motorised wings alone, the aerodynamic SmartBird was actually modelled on the herring gull. Made of lightweight foam and carbon fibre, powered by an electric motor inside its 'body', it can take off, fly and land all on its tod.

Unfortunately, pretty creations like SmartBird are totally impractical for anything other than mere aesthetic pleasure. There are plenty of videos on the world-wide-inter-web of similarly strange-looking contraptions flapping gracefully around the insides of warehouses and the halls of flash office spaces. But take it outside and one big gust of wind is all it would take to send the SmartBird careering off into the nearest tree, giving us painful flashbacks of losing many a Frisbee to the high branches.

Is there a solution? Step forward the latest wonder material: *dielectric elastomers*. A recent collaboration between engineers in China and the USA resulted in the discovery of this previously unknown property of a material. It's already really useful in robotics because it is very lightweight yet can be used to build motors and actuators.

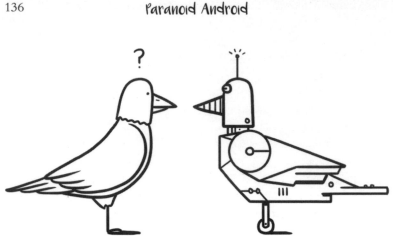

And as any roboticist will tell you, one of the major problems with building robots is that the more things you want them to be able to do, the more components you need to add. Very quickly, this can tot up to a robot that struggles to move under its own weight, smashing paving stones anytime it takes a stroll down the road. Robotic joints fashioned from dielectric elastomers come in very handy for making soft, flexible robotic components because they are able to create movement while remaining soft and lightweight. This makes them suitable for robotic hands, soft robots and shape-shifting robots, with which humans can interact without hurting themselves. And now this means we'll soon be able to construct flapping robotic wings that can go outside. Hoorah!

Robot wings constructed with dielectric elastomers would be different from their predecessors because, unlike earlier designs such as SmartBird, the energy conversion is much more efficient – even more so than real bird wings. This means they *should* be able to generate the kind of forces necessary to compensate for pesky gusts of wind, staying on course in all weathers. We don't know about you but as soon as they get these robot birds flying off the production lines we're gonna put in an order for a pair of hummingbirds. Ideally we'd like them to make it so the robotic flapping wings can eventually be attached to our own bodies meaning we can cruise about, in the words of Lynard Skynard, as 'free as a bird'.

GEEK CORNER: How exactly do you build yourself a flapping robot wing? You start by sandwiching several pieces of the soft, flexible elastomer between two slices of metallic electrode. Applying a high-voltage, alternating current across a rotary joint made of this dielectric elastomer makes it bend first one way and then the other, just like a flapping wing. But the alternating current used here is not like that produced by your car alternator: square-wave alternating currents 'alternate' in the sense that a high voltage – in the range of kilovolts – switches on and off, over and over again, to continuously manipulate the bending angle of the joint.

PERFECT ROBOT UPDATE: Flapping-winged robots definitely tick a box in our ideal-bot checklist, but a little closer to Earth there are some practical things we might also wish to consider, like how a robot could help keep us in tip-top condition.

Mirror, Mirror on the Wall, Who's the Sickest of Them All?

It's easy to be a hypochondriac these days: TV adverts, Google searches and shock-documentaries can all encourage us to worry, self-diagnose and convince ourselves that we have the plague. Wouldn't it be lovely if we had our own 24/7 health checker at home, one that lets us get on with our daily life worry-free and only alerting us when it's time to call in the doctor?

Thank goodness, then, for a new invention that helps us keep track of changes in our health before it gets really ugly. Laboratories in seven different European Union countries got together to create a 'magic' mirror that, while it wouldn't dream of passing judgement on 'who's the fairest of them all', could nonetheless give valuable feedback on changes in health that might save your life.

CHIC FACT: Sit in front of a mirror in a darkened room, about three feet away from it, and gaze at your reflection for about ten minutes. You should start to experience strange hallucinations, such as your face looking like it's made from wax or, even more dramatically, that it morphs into a monster's face.

Here's how it works: facial recognition software incorporated into the mirror looks for tell-tale signs in a person's appearance to establish their levels of stress and anxiety; 3D scanners keep track of weight changes according to subtle differences in face shape; multispectral scanners measure heart rate and haemoglobin levels; and gas sensors even analyse their breath to establish how much they have been drinking and smoking.

GEEK CORNER: Mirrors have been used for all sorts of medical purposes; for example, they can be used to ease the discomfort of phantom limb pain, which often develops after amputation. The loss of feedback from tissues in the missing body part often leads to the sensation that it is locked in a painfully contorted position. Simply placing a mirror so that the reflection of the surviving forearm overlaps with the position of the missing one and the visual illusion that the person *can* control the missing phantom limb enables significant pain reduction. Seeing the hand, wrist and fingers moving in the reflection enables the phantom limb to be manouevered into a more comfortable position, reducing pain by taking advantage of our brains' intrinsic plasticity.

By taking such measurements on a daily basis, the Wize Mirror can help a person keep track of their health and alert them if they need to get a medical check-up. Several illnesses, like heart disease and diabetes, can take an awfully long time to develop, yet once they do they can have a serious and irreversibly negative impact on that person's quality of life. So, heading these diseases off at the pass, i.e. before they get serious in the first place, could dramatically change, and ultimately even save, many people's lives.

CHIC FACT: Some animals recognise themselves in the mirror, as humans do after the age of around twenty-four months. Killer whales, magpies, gorillas, chimps, elephants and dolphins all pass the mirror-self-recognition test with flying colours.

The mirror is not actually a necessary part of this contraption, but it is very nifty. You see, they chose to embed all this kit in a mirror mainly because, vain creatures that we are, it's the one item we can usually be relied upon to stand in front of every day. We humans simply don't notice the subtle physical changes that occur in our bodies from one day to the next; it's only when we look at an old photo that we might notice that our face has got fatter, or thinner, or paler, or darker. But by then our health might have gone beyond the point of no return. The great thing about this technology is that it can keep track of those changes for us and give us the opportunity to tweak our habits before it's too late. Plus, having a mirror that checks our health gives us the perfect excuse to keep checking our hair.

PERFECT ROBOT UPDATE: Having a mirror to help keep tabs on our health would certainly come in very handy. So we'll be sure to integrate one of these into our dream robot. But no matter how much our medical mirror tries to keep us healthy, accidents will still happen. Hurt ourselves badly and we can find ourselves having to go under the knife. Having a surgeon you trust

is important, especially if they're digging around in vital organs like the heart and brain, so how would you feel if yours was a robot? Would you be nervous about having a non-human bot slicing away to its tin-heart's content – or secure knowing that the element of human error was removed?

Robotic Surgeons

A few years back, Dr Jack was asked to present four short films aimed at school kids to glorify the name of science and in particular engineering.* The topic of one film was an incredible device that gave human surgeons robotic superpowers via something known as the da Vinci Surgical System.

> **GEEK CORNER:** Google have teamed up with the multi-national pharmaceutical company Johnson & Johnson to develop artificial intelligence-enhanced surgical systems that can highlight blood vessels, nerve cells *and* tumour margins that are all extremely difficult to see with the naked eye.

Historically, surgery was always performed by a barber† because the barber shop was the place to go if you wanted tools sharp and (more or less) clean enough to chop through bone or hack into flesh, adding possible weight to the theory that Jack the Ripper may have been a barber (see p.35). As the years rolled by, surgeons took over and gradually started learning one very simple but important lesson: the bigger the hole you make, the more often you lose a patient. By continued refinement of surgeons' techniques, the holes through which they performed their surgery gradually became

---

* Admittedly, having a neuroscientist present something that aimed to inspire young people into engineering careers might be considered a bit of a fudge, but the Institute of Engineering and Technology figured that, without all those amazing engineers who build and maintain MRI scanners, Jack would never have been able to peer inside over a hundred human brains during his research career.

† That's why when doctors finally qualify as surgeons their title reverts from Dr to Mr, Mrs or Ms.

smaller and smaller. Eventually, fast-forwarding to modern times, they reached the point where, even with all the latest cutting-edge scalpels and retractors available to them, they could simply go no smaller.

Then, at the turn of the twenty-first century, the robotic da Vinci Surgical System came into play, allowing surgeons to manipulate its tiny little robo-fingers via remote control to perform surgery through several tiny incisions, meaning less blood loss, less pain and less time in hospital. Advances in optics then enabled surgeons to navigate in 3D. Some systems even use augmented reality to go beyond human visual capabilities to clearly display the borders between healthy and diseased tissue. Plus, the scars from the keyhole-sized incisions are so minuscule that no one could even tell you'd been on the operating table the next time you hit the beach. Happy days.

CHIC FACT: Robot-assisted surgery usually involves the surgical system being operated via remote control from a console sitting next to the operating table. As the system is operated remotely, this means that the surgeon *could* perform the surgery without even being in the same room, or even the same country, as the patient. If it was us on the slab we'd want a pretty cast-iron Wi-Fi connection!

However, there is a downside, because the next step would be to do away with human surgeons all together. With robot surgeons and pigeon pathologists on the rise (see pp.107–8) the days of skilled and hardworking *human* medical staff could soon be numbered and we're not sure about you, but we'd miss the human touch.

CHIC FACT: The da Vinci Surgical System, developed by Imperial College London, costs well in excess of $1,000,000 and to date has been used in surgery on the heart, pancreas, liver, lung, colon, brain, genitalia, kidney, bladder and bones.

PERFECT ROBOT UPDATE: Slowly but surely we're achieving our goal of creating the ultimate robo-partner. How great would it be if our swarming, flying, robot friend could also perform life-saving surgery on you. and your friends?! But medical care isn't all about operations and health checks. Sometimes it's simply about looking out for us.

Carer Robots

In the rare moments when we are brutally honest with ourselves, we know it's likely that we'll probably end up spending our twilight years living alone or in an old folks' home, given the ageing population thing.* Which sucks! As if that's not depressing enough, the problems stemming from loneliness and isolation in old age are more visible than ever, with several studies showing that feeling connected to other people is the number-one factor predicating good physical and mental health.

Robots may not be the obvious choice for care and companionship in our old age, but many companies are developing bots that can fill in the gaps when family, friends and social care are lacking (we're now reaching for the phone book thinking 'I really must reconnect with everyone I've ever met').

---

* Comparing birth rates to death rates in developed world countries, there is a steady increase in the proportion of elderly versus younger people, because medical advancement is keeping people alive for much longer and contraception/socioeconomic pressures are encouraging young people to have fewer children than in previous centuries. Consequently, there will be fewer younger people to look after the older generation than ever before in human history.

CHIC FACT: Robear – yup, it is a robot that looks like a bear – can gently pick up a frail adult and lift them to their bed or chair. There's also Paro the robot seal who has been developed in the UK for the NHS to provide company for people with dementia.

One project named ACCOMPANY* has been testing Care-O-Bots at several locations in the UK, the Netherlands and France, to provide physical and emotional support to elderly people. This may one day supplement – though hopefully not replace entirely – the care provided by family and human carers. Using machine learning and other forms of artificial intelligence, the Care-O-Bot could, for example, analyse images of a person's body posture and facial expressions to establish their mood, in order to suggest suitable activities or other types of stimulation that might improve that person's emotional state.

'Should I call a friend for you?'
'Fancy watching a film?'
'Shall we go for a nice walk in the park?'
'Would you like a foot rub?'

The Care-O-Bot can even help with menial tasks, such as cleaning, fetching food and drink or simply bringing you the remote control. It can also provide physical support in moving around the house, remind you to take your medication and call for help in an emergency. Or it could simply keep you entertained: playing games, telling you stories, turning on your favourite tunes and chatting away.

It's essentially a personal rent-a-mate and butler mixed into one. Thinking about it – why wait until retirement? We want a Care-O-Bot right NOW!

---

* ACCOMPANY stands (very roughly) for the Acceptable Robotics Companions for Ageing Years – in fact, we're surprised that they got away with that acronym.

PERFECT ROBOT UPDATE: Our dream robot must have these caring skills and we'd certainly want it to play a mean game of Jenga, too. But while robots playing functional, pastoral or even academic roles is a prospect that sits neatly within our comfort zones, the idea that robots could one day create works of art seems to get a lot of people's goats, but how close are we to creating truly artistic robots?

Artistic Robots

As the line between humans and robots becomes increasingly blurred it makes us wonder: what really separates us from machines? If robots can do so many of the things that we do, then what actually is it that makes human beings unique?

Some would say it's our souls and imaginations, expressed through our ability to create works of art. From Picasso to Beethoven, Shakespeare to The Beatles, the human ability to make art is special and unparalleled. Or so we thought . . .

Many roboticists have been busy developing robots with no real functional purpose other than to create interesting, entertaining, beautiful or original works of art, music and literature. And the results are scarily impressive.

Art

When one half of your authoring duo was asked to present a segment for a TV show on culture, she jumped at the chance to investigate the surreal world of robot visual artists.

While many robots out there can be pre-programmed to copy or create art, one stood out for being touted as being able to create unique works of art completely of its own volition. Paul – a robo-artist created by French artist and roboticist Patrick Tresset – was said to autonomously draw portraits inspired by its own artistic interpretation of its environment. Lliana decided to put 'him' to the test.

It was certainly a surreal experience, partly because Paul is actually *five* Pauls, each looking like a mechanical prosthetic arm with a biro on the end and a little camera on the top, checking you out. Having five little eyes giving you the beady while drawing you with its mechanical arms was, she imagined, quite a different experience from having Picasso paint your portrait in the South of France in the swinging sixties. Nonetheless, the five drawings Paul created were pretty impressive. A combination of facial recognition technology, a bunch of complicated algorithms and impressive computational power allowed each of Paul's arms to simultaneously interpret Lliana's face in a different way. The result was five unique sketches, each different from the other and each with its own merit. Each Paul had his own way of seeing the world. Far from a direct or accurate interpretation of what was sitting before him, each could change and develop its style over time. Dare we say that not only did each Paul have its own artistic view of the world, but perhaps even its own unique personality.

Impressive, indeed, but Lliana decided to go one step further. She took Paul's drawings onto the streets of London wondering whether people would be fooled into believing they were created by a human. Would they smell a mechanical rat? And how would it compare to a pretty lousy self-portrait Lliana drew of herself which, although lacking in Paul's technical ability, at least benefited from having that elusive human touch? The investigation was limited, with only ten random people asked to take part, but the results were certainly interesting and provided much food for thought: no one clocked that Paul's drawing was robot-created, but when asked to choose between Paul or Lliana's masterpieces eight out of ten preferred the human version. Perhaps there really is something intangible about the human spirit that can be translated through our artistic creations and never properly realised by robots, no matter how sophisticated or highly developed they become. Or maybe it's just a matter of time before the Tate will be hosting Paul's successor's first exhibition.

> **CHIC FACT:** ArtBots has been holding robotic art exhibitions featuring the work of robo-artists from around the world since 2002.

Music

Daft Punk aren't the only sound of robot rockers. Toa Mata, the LEGO robot covers band, bash away on tiny musical instruments to recreate songs by Depeche Mode or Kraftwerk when hooked up to an iPad. Their larger-bodied friends Robotic Church are themselves works of art – human-sized creations of twisted metal, wires and musical instruments. They play their own bodies, strumming away on string ribs and dancing with wild abandon. Meanwhile, German company Festo has created robots capable of composing unique musical pieces. Their robotic string quartet listens to pieces of music, before software breaks it down into its composite pieces. Pre-programmed compositional algorithms recreate something completely new, which is fed back to the band who 'listen' to it carefully. The band then reinterpret it and listen to each other as they play, creating a totally novel improvisational jam.

> **CHIC FACT:** The magazine *Nature* predicts that 90 per cent of journalism will be written by robots or computers by 2030. In fact, how do you even know that this book isn't written by one – do we even know? Are we real? HELP!

Literature and Poetry

In 1965 the eccentric, brilliant artist and mechanic Bruce Lacey was asked to take part in a poetry gathering at the Royal Albert Hall in London. Instead of sticking to convention and performing himself, he built a radio-controlled robot named John Silent who rolled onto the stage before belching and farting. His not-so-stuffy peers hailed it as a success, the height of avant-gardeism and a brilliant comment on poetry itself. John Silent didn't create his own unique works of poetry, of course, but others since then *have* demonstrated literary prowess of their own. Companies such as Narrative Science, for example, have developed algorithms that can turn data into journalistic stories and predict that the outputs generated by their program will win a Pulitzer Prize within the next five years. Perhaps it's only a matter of time before they develop robots with writers' block and a fear of deadlines ...

For the time being artists of the world can rest easy: we are some way from replacing them with robots. Good thing too: we certainly don't want to live in a world without any spontaneous, imperfect, and yet very *human* masterpieces of art.

PERFECT ROBOT UPDATE: Would we incorporate artistry into our perfect robot? Perhaps. We'd certainly enjoy one that could bash out a Rolling Stones tune at the drop of a hat. As long as they don't replace their human counterparts we're happy to have them, but how would we feel if they looked like us, too? Would that be unsettling? Or would it place us at greater ease?

Rise of the Android (Robots That Look Just. Like. Us.)

We really are a bunch of narcissists. Not content with having robots that can make stuff for us, look after us, operate on us, perform for us and work for us, now we want them to look like us, too.

The idea of androids is nothing new; the word originates from the early eighteenth century and literally translates as 'man-like'. It was used way back in 1863 to describe tiny toy robots and went on to frequent the imaginations of many great science fiction writers. From the French novel *The Future Eye* written in 1863, to 2015's *Ex Machina*, *Total Recall*, *Star Trek* and everything in between, androids have long been a futuristic fantasy confined strictly to our books, screens and dreams.

No longer! These days the question posed by Philip K. Dick in his seminal 1968 sci-fi novel *Do Androids Dream of Electric Sheep?* can now finally be answered. And it's all thanks to a fella named Professor Hiroshi Ishiguro (or, as we like to think of him, the 'Father of Androids'). In 2006 he created Geminoid,* which

---

* Gemini = the twins. Geminoid is an android built to look identical to Prof. Ishiguro. So you can see why he thought it would make a great name, right?!

entered *The Guinness Book of World Records* as the World's First Android Avatar, before deciding to improve upon this accomplishment a few years later with the beautiful Erica. Having had the privilege of meeting both the good professor *and* his creations, we're pretty sure he's going to get the job of making our robot feel more human.

Are you ready to get to know our two favourite androids? Ladies and gentlemen: we present Geminoid and Erica.

Geminoid

If we're to be completely honest about it, Geminoid looks a bit like he's had a stroke.* But from Prof. Ishiguro's perspective he made life much easier in a way that would get maximum respect from anybody who appreciates a cheeky lie-in. Why? Well, his main office is based at a university in Osaka, but his robots are all an hour's commute away in Kyoto. He regularly gives lectures, has meetings and holds tutorials at both sites and after a while the commute was becoming a real bore. So what did he do? He designed Geminoid in his own image and it genuinely *does* look very like him. Geminoid is controlled remotely by a person wearing special headphones fitted with movement detectors who's sat in front of a webcam equipped with special software to track their facial expressions – it doesn't even need to be Prof. Ishiguro.

Anytime Prof. Ishiguro couldn't be arsed to make his way to a lecture in person he'd get one of his staff to wheel Geminoid into the hall in Kyoto and then operate it remotely from his office twenty-six miles away in Osaka. When Ishiguro opens and closes his mouth, Geminoid opens and closes his. When Ishiguro speaks, his voice comes out of a speaker in Geminoid's mouth. When Ishiguro turns his head, Geminoid does the same.

What Ishiguro has achieved with this creation is virtual embodiment: he is able to create a physical, social presence in a roomful of people many miles away. Goodbye long and boring commutes and hello meetings, lectures, bank appointments (and weddings?!) from the comfort of your living room. Sadly, Geminoid can't go to the dentist for you – yet.

---

* He really doesn't have the most expressive face in the first place.

Now you're familiar with Geminoid we'd like to introduce you to Prof. Ishiguro's newest creation.

Erica

Erica is hot. There's no two ways about it. She is dead sexy. But if you look at YouTube videos of her you'd probably think we're being daft, desperate or maybe a bit of both. She's simply not that impressive until you get in a room with her. Erica is a much more sophisticated piece of technology than Geminoid. She can do certain things that we humans take completely for granted when interacting with other people, but only notice when they're absent. She moves her eyes to maintain eye-contact as you move your head, for example, which is something we humans do without even realising. Or making the tiny involuntary micro-movements we all do – no human can stand absolutely, completely stock still (well, apart possibly from those strange painted live-statue people who hang out in city centres and only move when you put money in their boxes). Both those things have to be specifically programmed into robots to make them appear more human-like. We take these things completely for granted and rarely think about them when communicating with other humans, but an android lacking these key features can be quite disconcerting.

Professor Ishiguro and the rest of the team behind Erica incorporated an array of Kinect XBOX movement trackers in the ceiling above her and a pair of bespoke 360° arrays of microphones to continuously keep track of where people and objects are in space. This enables Erica to make natural-looking head and eye movements to keep her gaze pointing in your direction throughout the interaction. Take it from us, it is disarmingly freaky when experienced first-hand.

Conversationally, Erica's also very impressive. Although the chat

Jack had with her was rudimentary, the ability to speak English *had* only been hacked into her repertoire a few days before his arrival. It's her capabilities in Japanese that are truly remarkable. She can understand simple conversational phrases and give appropriate, natural-sounding responses. Her artificially intelligent brain is now so sophisticated that she can conduct a genuinely fluent conversation, including all the correct vocal intonations and emotional overtones expected of a regular human conversation partner. Which is great, because the ultimate aim of these projects is to provide android companionship for the rapidly expanding elderly population in Japan. Although we're pretty sure some dirty dogs out there might see Erica, Geminoid and all the other incredible androids around as more than just friends . . .

Sex Robots

> Got to do my best to please her, just cos she's a livin' doll.
>
> 'Livin' Doll', *Cliff Richard*

The idea of humans feeling romantically inclined towards their creations isn't really all that new. Ovid's ancient Greek tale of Pygmalion, the story of an artist who falls in love with his own ivory sculpture, and more recently science fiction films such as *Blade Runner*, *Austin Powers*, *Ex Machina* and *Westworld*, all present us with their interpretation of the physically perfect sex-bots. But these fantasies are no longer confined to the imaginings of writers and artists. Sex robots have now become a thing of reality, with some even predicting that human–robot sex will be more common that human–human loving by the middle of this century.

One of the most popular versions is the walking, talking, shag-android Roxxxy – available to all you robot-loving punters for the hefty price of £5,000. Essentially, Roxxxy is a very advanced computerised sex doll – 'she' has silicone 'skin' over a rigid skeleton, with voice recognition and speech-synthesis technology. Roxxxy can chat, display emotions and, importantly, have sex for as long as her batteries last, which – for those of you interested – is an eye-watering three hours. She even comes with five different 'personalities', with horrifically crude names including Frigid Farrah and Wild Wendy.

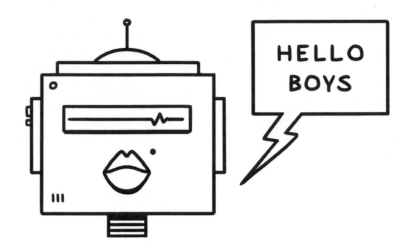

Unsurprisingly, quite a few feminist writers (male and female) have had a few choice words to say about these delightful nicknames, as well as about the sex-droids themselves.

However, when it comes to realism, manufacturers Actroid–DER are leading the way. Their humanoid robots are so lifelike that some claim they can't even tell them apart from a real person. While its makers remain coy about the full potential uses of their rentable female bots, describing them with lines like '[she's] even cuter than her older sister!' or 'her astonishingly small face is capable of creating exotic facial expression', there are hints that their designated function may involve a bit more than polite conversation.

Meanwhile the makers of the incredibly lifelike silicon and electronic maid Aiko claim that having sex with her isn't their primary objective – she also cleans up, keeps you company and cleans your ears with cotton buds (you lucky thing!) – they don't deny that it's a distinct possibility.

What, then, are the benefits of this new bot-humping trend, which will no doubt continue to grow in popularity as time and AI developments move on? On the one hand, robots are clean, disease-free and could eventually replace the need for the highly exploitative sex industry. The advantages of this could be huge. Robots (unlike humans) won't be traumatised by even the kinkiest desires of the most depraved sadists, so using them could reduce the incidence of human suffering across the world *and* they can't be murdered by any prostitute-hating psychopaths. The ripple effect of a perfectly legal sex robot industry would also be a crippling impact on the criminal underbelly of our cities. And for married people it could offer a more ethical option to having an affair. Plus, with so many isolated people out there in need of some TLC what's a little loving from a machine between lonely hearts? Besides, fans of sex-bots may also ask, are they really that different from already broadly accepted sex toys?

On the other hand, many are concerned about the ethical implications of a blooming sex-bot industry and are calling for an outright ban. They fear it could reinforce an outdated and dangerous view of women as existing purely for male gratification, valued for their physical attributes alone (we wonder how the observers of Frigid Farrah came up with that one?!). It could also have a serious impact on our ability to form real, meaningful relationships; with robot sex on tap to satisfy our immediate desires, men and women may no longer have any real need to overcome their fears/natural laziness to engage in genuine human conversation and courtship. Think that sounds crazy? Well, Facebook generation, we bet our parents would have similarly thought they'd never see a day when so much of our social interaction would happen on computers either.

A future filled with human sex-bots isn't the only thing on the table for debate. One artificial intelligence researcher believes that love and even marriage between robots and humans will be taking place by 2050. After that, who knows? Human–robot couple therapy, affairs, divorce and crimes of passion?

Whether it's a good thing for humanity, or the beginning of the end of one of the most important aspects of human interaction, one thing's for sure, a robot won't roll you over into the wet patch, light up a post-coital cigarette and warn you not to let the door hit you on your way out. Unless that's your kind of thing, of course.

PERFECT ROBOT UPDATE: Um, OK, though it would be fine if our dream robot was relatively attractive, call us old-fashioned, we're not sure we're ready to jump into bed with one quite yet. However, when it comes to cyborgs – that is, humans who've had certain parts of their bodies replaced by robotic components – we're great fans of a bit of that.

Robotic People – Real-Life Cyborgs

Cyborgs were once entirely fictional people who could extend their abilities beyond the normal human range using technology and mechanical parts. Star-studded celebrity cyborgs of screens large and small include Robocop, Darth Vader and good old Inspector Gadget. But these days cyborgs are becoming a reality as people implant technology in their bodies and brains for all sorts of reasons, from medical to pure convenience. Here's a smorgasbord of some of our favourite real-life cyborgs.

Abracadabra

American Amal Graafstra got so hacked off with having to take out his swipe card to get through security doors that he decided to embed microchips into the webbing between his thumb and forefinger, just so that he could wave doors open like some kind of Jedi master instead. At 2 × 12 millimetres these cylindrical chips are about the size of a grain of rice and work using radio frequency identity technology, aka RFID. Amaal didn't stop at doors. He also uses his magic hands to unlock his computer and his motorbike. Personally, we'd have been tempted to adopt the simpler approach of reprogramming the offending doors to respond to the magic words 'open sesame!'. But, hey, each to their own and all that.

Zim Zam Zoom

While many a one-eyed seafaring pirate has opted for the classic patch-look, one-eyed film-maker Rob Spence – being a landlubber and all – navigated a different route through the seas of monocularity. He thought it would be cool to have a miniature digital video camera fixed into his empty eye socket so he could shoot film that gave a genuinely first-person point-of-view perspective on the world. With the help of some incredible engineers and medical health practitioners he became Eye-Borg man! Although it's not actually plugged into his brain, he nonetheless controls the camera angle through spontaneous eye movements. Not only can he record whatever he decides to look at, totally hands-free, but the resulting footage even has the neat feature of capturing his natural eye blinks.

Alakazam!

Neil Harbisson was born completely colour blind. As an artist he became understandably frustrated by his inability to see colour and decided to become a cyborg to sort it out.

He suspended a digital camera in mid-air just in front of his fore-head like a third eye, which would cunningly transduce whichever colours came into its line of view – which Neil controls simply by turning his head in the desired direction – into different types of vibrations at the back of his skull. The vibrations travel from the back to the sides of his skull where they are detected by the inner ear, which is usually tasked with detecting sounds coming in through one's ears. The frequency and amplitude of the vibrations change according the colour detected and his brain has miraculously learned to use this signal as the basis for his perception of colour. Pretty mind-blowing.

Neil first came to our attention when a story about him hit the newspapers – the British passport office was insisting that he would have to remove his third eye for his official passport photo. He objected, arguing that the tech was now a part of his body and that he wouldn't feel complete without it. He added that his robo-eye was genuinely such a fundamental part of his identity that he now even dreams in colour, and that he never, ever takes it off. It threw up all sorts of interesting ethical questions. Neil eventually won the battle at which point the millions of people around the world who've had passport applications returned for pedantic reasons col-lectively reached out their metaphorical arms to give him a giant hug of support.

CHIC FACT: In the future Neil Harbisson plans to develop the technology further so that he can detect not only colour but also parts of the electromagnetic spectrum that humans can't normally see, like infrared (for night-vision) and ultraviolet (for bug-, snake- and turtle-vision) wavelengths. Where do we sign up?!

GEEK CORNER: Transduction is the process whereby a given impulse is converted from one form of energy to another; in Neil Harbisson's case, from light to vibration. In the back of the eyeball, on the other hand, light is transduced into electrical pulses that the brain can work with instead. The transduction technique of converting visual information into vibrations has also been exploited to enable blind people to go rock climbing by converting visual-shape information into vibrations on a special tongue pad.

PERFECT ROBOT UPDATE: What does this mean for our dream robot? For starters, we've decided it's going to have a special camera embedded in one of its eyes, one that can see in wavelengths invisible to the human eye, although we're going to pass on surgically attaching a vibrating wire to our skulls because, quite frankly, we like wearing hats too much. However, when it comes to combining robo-tech with our own human bodies in true cyborg fashion, there's one lady who's really shown us how it's done . . .

The Woman Who Flew a Plane with Her Mind

There are few abilities more aspirational than being able to fly a plane. Angelina Jolie, Bruce Dickinson and John Travolta can all do it, and, damn it, fifty-five-year-old businesswoman Jan Scheuermann wanted to do it, too. The only catch was that she happens to be quadriplegic – paralysed from the neck down since 2003 as a result of a rare genetic disease. But Jan wasn't going to let a small obstacle like near-full-body-paralysis stop her from doing extraordinary things and neither was the team of scientists she was working with in the Pentagon's experimental robotics programme.

Jan had been selected to work with the US Defence Advanced Research Projects Agency (DARPA) prosthetics team back in 2012. First she gave permission for surgeons to implant electrodes directly into her brain – the left motor cortex to be precise, the part of the brain that controls movement in the right side of our bodies. This procedure allowed her to control a robotic arm, which she nicknamed 'Hector', simply by imagining performing various movements with parts of her paralysed body. After a few months of practice Jan had full control of Hector, commanding him to pick things up, put them down and even feed her chocolate using the power of thought alone – oh, for more Hectors in the world!

The surprising thing was that Jan seemed to have an extra-special ability – as if feeding yourself chocolate using a robotic arm named

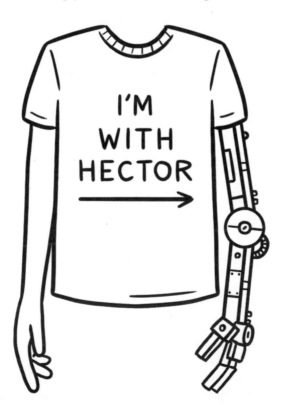

Hector just by thinking about it wasn't special enough. Jan could control robotic arms on both the left *and* right sides using the electrodes plugged into her left motor cortex. Usually we can only control the right side of our bodies with our left motor cortex and vice versa. Realising that Jan was a pretty extraordinary individual, the research team felt she was ready for a new challenge: flying the Pentagon's latest fighter jet . . . in a simulator.

CHIC FACT: While research like this could revolutionise the quality of life for people with disabilities, imagine controlling every aspect of a smart home through the power of thought alone; however, there are also much more potentially sinister applications. Military powers and, come to think about it, anyone with sufficient cash, will eventually be able to pull the trigger just by *thinking* about it – scary stuff!

We'll hold our hands up and admit that flying a simulator isn't *quite* the same thing as flying a real plane, so we may have slightly misled you with the story title, *but* simulators are what all pilots train on initially. Apart from safety risks there is virtually no difference in the skill or ability required. And, *hello!* Jan is paralysed, so this is still utterly amazing. Instead of using joysticks, like your average able-bodied pilot, Jan was able to control the digital F35 Joint Strike Fighter using the power of her thought alone. All it took was for the Pentagon scientists to unplug her from Hector and plug her into the flight simulator instead.

In the journey from feeding herself chocolate to using her brain electrodes to fly one of the most sophisticated aircraft in today's skies,

Jan proved that if you really put your mind to something, nothing can stop you.

CHIC FACT: Jan wasn't the first person to fly a plane with her mind. In 2013 a remote-controlled helicopter was flown through hoops using the power of thought. Then, in 2014, a group of German scientists got seven pilots to control a flight simulator by thought alone – but, unlike Jan, who has electrodes implanted into her brain, these participants instead wore electroencephalography (EEG) caps, which hold dozens of electrodes against their scalp in order to detect brain activity across the skull. A bit tame for our liking!

The Final Perfect Robot

There you have it: the perfect robot. By navigating the latest robot-related breakthroughs we have successfully managed to paint a picture of our very own perfect robot. An attractive, chatty, artistically talented, flapping, caring, life-saving, mind-controlled android with a backpack full of tiny little robo-roaches in case of emergencies.

But the question remains: if our dream robots can paint like us, care likes us, chat like us, look like us, screw like us, save lives like us and generally be totally like us, then at what point does that render us completely obsolete?

The advancement of robotics will continue to pose huge philosophical, moral and ethical quandaries, ones that we, and generations to come, will have to carefully try to navigate. But there's no point fighting it. This chapter has clearly shown us that, in the words that we began with, the robots really *are* coming. They continue slowly but surely to infiltrate our everyday lives, but as to how much and to what extent we will come to rely upon them, that remains to be seen.

6

Let's Talk About Sex, Baby

Talking about sex – tricky, isn't it? We are obsessed yet embarrassed by it. Enthralled and appalled; thrilled and disgusted. There's no other subject that makes us feel such a heady mixture of conflicting emotions: awkward, excited, shy, jealous, flushed, ashamed, ecstatic, depressed, horny . . .

Talking about it seems no less confusing than doing it. The mere mention of sex can make middle-aged men blush or teenagers shriek. Yet we think about it frequently – up to fifty-five times a day.* How is it that we can be embarrassed about discussing it with our own parents, knowing full well that it's the very thing they got up to in order to create us in the first place? How can such a taboo subject be the very key to the survival of our own species?

It's a conundrum, but perhaps it's this great dichotomy between our simultaneous awkwardness and fixation with sex that can make

---

* Figures on how many times people think about sex every day vary wildly depending on methodology and how you define 'thinking about sex'. This figure is based on a study that asked participants to record sexual thoughts on a tally counter over the course of a week. The maximum recorded was 388 times a week, which was divided by seven to get the figure of a maximum of fifty-five times a day.

our ideas, habits, myths and practices of it so strange, bizarre and downright freaky.

In the following pages we'll be forgetting all the niceties and getting right down and dirty. We'll expose some of the weirdest historical beliefs, the oddest recent research revelations and the kinkiest habits of man (or woman) and beast alike. Strap on, readers: things are about to get a little hot under the collar.

The Weird History of Sex and Science

The quest for truly great sex can be a lifelong journey and one that can take patience and dedication (just ask Sting). Today, there's a plethora of aids, medications, techniques and surgeries you can invest in to enhance your sex life, and a whole host of ways to study the science of sex . . . but first, let's wind back the clock and find out what our ancestors got up to under the covers and why.

The Vagina Myths

The strange myths, beliefs and treatment of the female genitalia over the history of human life on Earth have varied from the ridiculous to the outright cruel. Through a combination of ignorance, misogyny and patriarchal dominance women's lady bits have been accused of some mighty absurd things, including:

Vagina Dentata: stories about vaginas lined with teeth are prevalent in many ancient folklores. Indigenous cultures from places as varied as Samoa, North America, Russia, India and Japan tell stories of women castrating men with a sexual chomp of their vaginal gnashers, or of men pulling teeth from their partners' privates pre-sex and even of women having meat-eating fish lying in wait inside their lady holes. Perhaps this was man's way of spreading the myth that women

are biologically penis-hating beasts, or maybe it was an ancient and rather effective way of convincing young men to keep it in their pants. However, interestingly – though most likely unrelated – in more recent times teeth *have* actually been found growing inside female reproductive organs. Dermoid cysts or tumours – containing all sorts of tissues such as hair, bones and up to three hundred teeth – have been observed growing on women's ovaries, but also in men's testes, on brains, necks and bladders. A terrifying prospect – one worthy of a Stephen King plot (*The Teeth Within*, perhaps?).

Vaginas are inside-out penises: the idea that women are simply inverted men dates back to the time of the ancient Greeks, when it was first suggested that vaginas are inside-out penises. Ovaries were also considered to be internal testes, the uterus an inner scrotum and labia underdeveloped foreskins. One seventeenth-century theory had it that the 'problem with women' was that they lacked heat and so tucked their reproductive organs inside themselves to keep them nice and toasty. However, should a woman ever get too angry, hot and bothered or unduly excited, her body temperature would rise and she risked spontaneously overheating and turning into a man. The likely lesson here: ladies, keep your cool, remain calm, orderly and controlled (by men). 'Calm down dear, or you might turn into a man.'

Period Power!

From getting cramps to forking out for sanitary wear, women already have it tough when it comes to getting periods. To make life really fun and interesting, various patriarchal societies decided to come up with a whole load of ridiculous ideas about periods being clear evidence that a woman is somehow dirty or unholy. The earliest writing about the magical powers of women's periods dates back to AD 77 when Pliny wrote that mere contact with a menstruating woman:

> ... turns new wine sour, makes crops wither, kills grafts, dries seeds in gardens, causes the fruit of trees to fall off, dims the bright surface of mirrors, dulls the edge of steel and the gleam of ivory, kills bees, rusts iron and bronze, and causes a horrible smell to fill the air.

Is that all, Mr Pliny? You may have left out 'and causes death to cute puppies'.

Sadly, even today the idea that menstruation somehow makes women unclean persists, with many religions banning females from a number of activities during their time of the month, including entering a place of worship, touching holy books, going into the kitchen, wearing flowers, talking loudly, having sex, driving a vehicle or mounting an elephant.

Sex Manuals – Baroque-Style

One of the earliest Western sex manuals, written in 1680, was called *The School of Venus*. Samuel Pepys described in his diary how he came across it, and though he thought it 'the most bawdy, lewd book' that he'd ever seen. He promptly bought himself a copy, read it from cover to cover and then burnt it in shame. The surprisingly forward-thinking book is written as a dialogue between a teenage girl and her older cousin who shares her knowledge on everything from orgasms, to friends-with-benefits and extra-marital affairs. Our wise lady also shares personal insights with her younger cousin regarding the male anatomy, describing how a penis – or 'Man's Yard' – hangs down 'like a cow's teat' and how a scrotum is 'like a purse' containing balls not too dissimilar to 'Spanish olives'. There's even a hefty, illustrated description of 'different positions' as the older cousin describes her own sex life with her husband, and shares that 'sometimes we do it sideways, sometimes kneeling, sometimes crossways, sometimes backwards … sometimes wheelbarrow, with one leg upon his shoulders, sometimes we do it on our feet, sometimes upon a stool'. Phew – those Baroque women didn't half know how to have some fun.

Hysteria in the Eighteenth Century

For the past two thousand years in European history the word 'hysteria' had zilch to do with The Beatles, One Direction or the Black Friday sales and everything to do with medicine. The 'medical condition' of hysteria, so the invariably male doctors believed, was an affliction of women who had had a disturbance in their uterus (the word itself 'hysteria' stems from the Greek word for uterus – *hysterika*). The symptoms, so they said, were irritability, anxiety, sleeplessness and sexual dysfunction. A woman diagnosed with hysteria could have a range of 'cures' prescribed for her. Anything from horse riding, to 'marital lust', to having doctors or midwives massage her genitalia with their hands or vibrating tools, and – prepare yourself for this – by spraying her vagina with water until she reached orgasm (or, as they called them, 'paroxysms'). All were all considered acceptable 'treatments'.

CHIC FACT: The earliest vibrators were steam-driven, and it's said they were often used to treat 'hysteria' by inducing 'paroxysms'. However, the electric vibrator itself started off its life not as a sex toy, but, rather, as a tool of medicine. Pioneered in the 1880s by Dr J. Mortimer Granville, it was initially used to relieve men of muscle pain.

Sex Surgery in the Swinging Twenties: Men

The 1920s saw a large handful of men and women go to extreme and disturbing lengths to improve their sexual vigour and there were none more bizarre than the monkey testicle implant craze.

GEEK CORNER: Testosterone is the main hormone produced in the testes and it affects sex drive, red blood cell production, bone density, body hair, muscle mass and sperm production. Many transgender men have testosterone therapy, as do those with a condition known as hypogonadism, a condition in which the male body doesn't produce enough testosterone, but they aren't the only ones. As all men age, their testosterone levels gradually decline by about 1 per cent a year from the age of around thirty-five, so many middle-aged blokes in search of everlasting youth are giving it a go, too. Results are mixed and side effects include sleep apnoea, brain blood clots and, surprisingly, enlarged moobs. Doh!

Men who approached a Russian doctor named Serge Voronoff for a little help in the sack were met with a rather radical solution. Voronoff was just a tad fixated by the power of testicles, firmly believing that they held the secret to age reversal and sexual prowess. Fellow physiologist Charles-Edouard Brown-Séquard was already making headway in this field and was busy eating monkey testes, or injecting patients (and himself) with a juice made from guinea pig and dog testicles and semen. Brown-Séquard self-reported a 'rejuvenated sexual prowess', possibly due to the placebo effect, and the scientific community dubbed it the 'Elixir of Life', greeting it with a mixture of excitement and ridicule.

Meanwhile, Voronoff, also inspired by his early studies on the effects of castration on eunuchs, was developing his own master plan. When wealthy men came to him for a cheeky little sex drive boost they would soon find themselves on the operating table, with a few thin slices of young healthy chimpanzee or baboon testicles being grafted onto the inner surface of their scrotums. He had previously used human men's testicles – usually taken from recently executed convicts – but after getting into a spot of bother with the law, Voronoff – who later became known as 'the monkey gland expert' – shifted to apes, even setting up his own ape farm in Italy to breed them.

Slicing primates' balls off and transplanting them onto men's scrotums sounds insane, cruel and incredibly painful for all involved and yet, after the first 'monkey gland transplant' took place in 1920, hundreds of men lined up to follow suit and by 1940 around two thousand operations had taken place all over the globe.

Voronoff's influence was felt in popular culture, too – in 1923 Arthur Conan Doyle wrote, in a Sherlock Holmes story, of an ageing professor who drinks monkey elixir to rejuvenate his flagging libido'.* Meanwhile, the poet e. e. Cummings wrote about the 'famous doctor who inserts monkey glands in millionaires', and the Marx Brothers sang 'if you're too old for dancing, get yourself a monkey gland'.

Remarkably, the scientific community was also initially impressed

---

* In Arthur Conan Doyle's 1923 tale 'The Adventure of the Creeping Man' the professor drinks the monkey juice, but then goes insane, turning into a sex-crazed apeman and scaring the bejesus out of Victorian ladies.

and in 1923 seven hundred leading surgeons at an international meeting declared Voronoff's work rejuvenating men's vigour a giant success. Sounds nuts? Maybe, but maybe not. Because, despite the science world eventually turning against Voronoff and ridiculing his work, in 1990 endocrinologists said that we should revisit his studies, because testicles are one of only a few organs (alongside eyes and the placenta) that don't readily reject donor tissues, thanks to special cells which create a barrier to the immune system.

Either way, there surely has to be an easier and more humane way to boost your sex drive – anyone for oysters?

Sex Surgery in the Swinging Twenties: Women

Before the women reading this start falling about laughing at these seemingly ludicrous attempts by men to boost their sex drives, you may be horrified to hear that Voronoff used his 'skills' in transplants on women, too. Yup, monkey ovaries were actually transplanted into women in an effort to 'rejuvenate' them. In fact, throughout history there have been examples of women going to absurd lengths to improve their capacity for sexual pleasure. One of the most extreme examples is that of Princess Marie Bonaparte, the great-grand-niece of Napoleon Bonaparte who, in the 1920s, started to wonder why she wasn't climaxing from penetration alone. She decided that the best way to sort it out would be to have the ligaments connecting her clitoris to her body snipped so that it could be relocated closer to the vagina. Because, you know, mutilation is always such a great idea when trying to make something more pleasurable. Unsurprisingly, the operation didn't work. But did Marie decide to call it quits and accept that the whole thing had been a monumental failure? Of course not! Instead, she decided to go under the knife again and move it closer still. Second time unlucky again, at which point Marie accepted that her dodgy experiments were a failure. Sadly, she had to live with the unpleasant results of self-experimentation for the rest of her life. And so ends the alternative story of *The Princess and Her Pea*.

While history describes many people who've gone to some extreme and painful lengths to boost their sex drive, thankfully today we have many more pleasant options. But how many traditional aphrodisiacs still doing the rounds today actually stand up to scientific scrutiny?

Do Aphrodisiacs Work?

Having good sex can be a tricky business, with over 43 per cent of women and 31 per cent of men reporting some sort of sexual dysfunction at some point in their lives. In such times, a little helping hand can sometimes go a long way and, for those of us who don't fancy popping pills, it's easy to see why natural aphrodisiacs are so appealing. But do they really work when put under the microscope?

Oysters

Have you ever sat at a romantic date, gazing into your loved one's eyes and asked yourself why on earth the slightly snotty, salty, gloopy-looking raw mollusc in front of you is supposed to be a turn-on? Could it be because Aphrodite, the Greek goddess of love, was said to have been born from the sea? Or was it the rumour that Casanova, the famous eighteenth-century lover, fuelled his insatiable libido by eating fifty oysters for breakfast every morning? Whatever the reason, oysters are today probably the most famous of aphrodisiacs, and several scientific studies have put them to the test.

One research team of American and Italian scientists analysed a group of shellfish named bivalve molluscs (including oysters, clams, cockles, mussels and scallops) to see if there was any substance to the long-standing fishy aphrodisiac belief. They reaffirmed what previous studies had found: that oysters have a particularly high zinc content, which *could* be significant given that zinc is found in sperm – each ejaculation contains around two milligrams. However, the most exciting part was what they discovered next – that two important amino acids (i.e. building blocks of proteins) were also present: D-aspartic acid and N-methyl-D-asparate.

The lead scientist behind the study – the aptly named Dr Fisher – thought this finding particularly interesting because these amino acids are rarely found elsewhere in nature. Earlier experiments by Dr Fisher's team had also shown that injecting rats with these particular amino acids led to a rise in sex hormones: testosterone in the males, oestrogen and progesterone in the females. So could it be that these amino acids are the key to the oyster's desire-inspiring qualities? Is the legend of oysters as aphrodisiacs really true? It's possible. And Dr Fisher's top tip is that if you want to slurp down on oysters with the highest concentration of those two key amino acids, eat them during their breeding season and always enjoy them raw.

However – and as any good scientist worth his or her salt will tell you – results from a rat-based study (such as Dr Fisher's) are all well and good, but until a randomised, controlled trial on *humans* shows a definite link, then, as far as they're concerned, the idea that oysters are aphrodisiacs are still nothing more than an old (fish) wives' tale.

Verdict: The jury's still out.

Chocolate

Mmmm ... chocolate. Almost certainly a way to your lover's heart, but is it the way to their erogenous zones, too? The association between chocolate and seduction goes way, way back: the Aztecs called it the 'nourishment of the gods', believing that their emperor Montezuma drank chocolate to boost his libido before he went to 'visit' his harem of wives. Ever since, chocolate has been a staple for any serious wooer's repertoire, but is there any actual scientific fact to back up the aphrodisiac rumours?

One doctor in the 1980s wrote a book called *The Chemistry of Love*, sparking what became known as the 'Chocolate Theory of Love'. He believed that the high concentration of the phenethylamine (PEA) in chocolate is what gives it its sexy powers. The theory relied on the fact that people who fall in love also show raised levels of this amphetamine-like neurotransmitter, which triggers the release of feel-good chemical dopamine and certain endorphins, suggesting that munching down on a bar of Dairy Milk could potentially mimic the feelings of love.

Since then, further studies have shown it's not *quite* so simple and that even guzzling several pounds of chocolate wouldn't cause a noticeable change in PEA levels in our bodies. However, many believe PEA could still have a role; for example, the taste or sight of chocolate could boost PEA in the brain. Other substances discovered in chocolate – such as methylxanthines, tyramine and cannabinoid-like fatty acids – have also been suggested as possible ingredients that give chocolate aphrodisiacal qualities.

However, despite these and many other possible candidates for chocolate's aphrodisiacal ingredient, a controlled study in 2006 found that once age was taken into account there was no difference in sex drive between women who ate a lot of chocolate and women who ate hardly any at all. Bummer.

Verdict: Myth (with possible placebo effects).

Rhino Horn

The most infuriating of all the aphrodisiac myths is that ground-up rhino horn can somehow boost your sex drive. This was a

popular belief in nineteenth-century Western culture (thousands of miles from the nearest rhino), primarily due to the phallic shape of the horn. It's all thanks to something known as the 'doctrine of signatures', which draws on a belief dating back to AD 40 that anything that looks like parts of our bodies can be used to treat it, because, you know, God was trying to give us handy clues and all that. So, walnuts were thought to be good for treating head ailments, yellow plants were perfect for treating urinary problems and rhino horn was good for anything ... rhino horn-shaped (ahem).

Thankfully, today most of us have caught onto the idea that 'God' (or nature) isn't quite as overt as all that and rhino horns are rarely used as an aphrodisiac any more. Sadly, ground up rhino horn is still thought to reduce fevers and other ailments in Eastern medicine and so conservationists have had to get seriously creative when it comes to protecting their horns and precious lives (see pp.221–3).

Verdict: Load of God-bothering codswallop.

Maca

Anyone who's been within a mile of a gym or within ear shot of a nutrition-obsessed buddy will have heard of maca powder (aka *Lepidium meyenii*), a high-protein Andean root vegetable often added to healthy shakes or snack bars. But it's nothing new; Peruvians have even been putting a cheeky dose in their farm animals' feeds since pre-Columbine times as a way to encourage breeding. Today, the pulverised tuber is thought to have all sorts of medicinal powers, including relieving hot flushes, boosting fertility and that all-important sex drive. But does the evidence stack up?

Tests on rats in 2000 proved to be pretty promising, with doses of maca increasing libido and erectile function, but, as we know from our oysterly investigations, people are not rats (well, not all of them, anyway) and a rodent study does not a human result make.

However, human studies have been pretty encouraging, too, with the key ingredients phytosterols and phytoestrogens put to the test. Many studies, using those all-important randomised, controlled trials, have indeed shown that maca can boost libido, help combat erectile issues and other sexual dysfunction.

Verdict: More research needed but this one certainly looks promising.*

Research will no doubt continue, and many other possible aphrodisiacs will be mooted, but meanwhile getting in the mood could boil down to something a whole lot simpler – the power of smell . . .

"LET'S TALK ABOUT SEX BABY!"

The Scent of Attraction

Ever smelled a dirty sock or a sweaty old T-shirt and thought: 'Cor, I fancy me a bit of that'? Possibly not consciously, but these stinky items may be sending powerful signals to our brains that the creator of the smell is one seriously sexy beast.

The idea that pheromones could attract us to others has been floating around for some time now. Kind of makes sense when you think how much we fork out on perfumes, body sprays and aftershaves that promise to make us utterly irresistible. Some species (like bees) use such airborne chemical signals as a form of communication, but can smell really be the key to sealing the deal in our own species?

---

* NB Geek Chic does not encourage or advocate the use of maca or any other supposed aphrodisiac. As with all natural supplements, you should consult your doctor before starting on any new regime – including sexy ones!

CHIC FACT: Aside from pheromones, people are also generally more attracted to those with symmetrical faces, supposedly because we subconsciously believe they'll have better genes.

One classic study asked forty-four men to wear a previously unworn T-shirt to bed for two nights in a row, washing only with unscented soap before bedtime. Forty-nine women were then asked to sniff the pongy, body odour-saturated T-shirts (lucky ladies!) and to rate the ones that smelled most attractive to them. The results showed that women consistently preferred the smell of men who were immunologically different from them, which makes sense. If you're attracted to someone who has an immune system that differs from yours genetically, then any potential future kids are more likely to have the best of both worlds, and be able to fight and survive a wider variety of diseases.

The study also showed that the smelly shirts the women preferred also reminded them of some of their past lovers, suggesting that who they chose to date was affected by smell in a consistent way. Even within close-knit religious communities – such as the Hutterites – who marry strictly within their small culture, they still somehow manage to find partners who are immunologically different from them. And there we were thinking our chosen lovers were simply sexy.

GEEK CORNER: Pheromones don't actually have a distinctive scent. However, when you sniff a piece of clothing worn by another person, these odourless pheromones are mixed in with other gaseous chemicals that give that person their characteristic body odour. In any given person, their scentless pheromones will always be mixed in with more or less the same profile of scented gases and so they come to be associated with each other. When people are asked to rate how attractive they find the overall smell, their conscious appraisal of the scented component is thought to be substantially influenced by their subconscious response to the pheromone.

The pheromone giveaway about how similar we are genetically to a prospective lover is called the major histocompatibility complex (MHC), of which there are four hundred different variations. Another pair of pheromonal sex culprits are two human steroids called androstadienone (found in male sweat) and estratetraenol (associated with women). Neither has a discernible smell, but sniffing androstadienone was found to enhance women's moods and make them judge men as more attractive, while in certain circumstances estratetraenol boosted men's moods and arousal. They were also both found to appeal differently to each gender according to their sexuality when inhaled. One study showed that sniffing androstadienone while looking at dots on a screen made to move like walking humans of ambiguous gender made heterosexual women and homosexual men judge the ambiguous figures as more masculine. Meanwhile, estratetraenol, the female pheromone, made heterosexual men (but not hetero-women) view the walking dot figure as more feminine – gay women and bisexuals showed mixed reactions. Looks like these natural pheromones can affect how masculine or feminine someone appears to be, depending on your gender and/or sexual orientation.

CHIC FACT: Men find the smell of tears a turn-off. A 2011 study found that men who smelled human tears were less aroused than when they smelled ordinary water and salt.

Nothing here is certain, however, with researchers still unsure if the two steroids tested really are human pheromones, after all. Complications arise from the fact that androstadienone can be found in women as well as men and that to date estratetraenol has only ever been found in the urine and placenta of pregnant women.

CHIC FACT: Testosterone may also play a key part in the scent of attraction as related to reproduction. Researchers found that during the most fertile times of their menstrual cycle women rated the smell of men with the highest levels of testosterone as the most attractive.

GEEK CORNER: The major histocompatibility complex (MHC) is a set of cell surface proteins, which recognise foreign molecules such as pathogens or donor cells and determine whether or not they are compatible with our bodies (and therefore how to treat them). MHC determines, for example, how compatible an organ donor is before transplant. When it comes to sexual selection, detection of air-borne MHC molecules allow our brain to analyse these pheromone signals to determine how genetically similar or dissimilar a person is to someone of the opposite sex. Under normal circumstances a heterosexual woman will have a preference for pheromones that signal a genetic dissimilarity. This is thought to be a mechanism that evolved to help prevent the conception of babies with genetic abnormalities that inevitably result from incestuous couplings and to create offspring with the strongest immune systems. Interestingly, in pregnant women this preference reverses, as the mother-to-be's priorities shift from finding an appropriate father to reinforcing bonds with family members who might assist with child-rearing activities.

Whatever the whiffy mechanism involved, many are so convinced that the key to true love and attraction is compatible pheromones that a smell dating service has recently been launched in New York. The first 'mail order dating service' matches people up depending on their scents. You sign up, wear the same T-shirt for three consecutive days with no deodorant, then send it back to the dating service. In return you get swatches of other people's pongy T-shirts to sniff. If you like someone's smell and they like yours, you've effectively swiped right and, wahey, it's a match.

Worrying about whether your pheromones are compatible is just one of the myriad things we Earth dwellers need to consider, but for those lucky enough to travel up into space there's a whole host of other issues waiting to greet them.

Sex in Space

In 2009 Professor Stephen Hawking announced that in order for the human race to survive we will have to find a way to successfully contend with the difficult environment of space. One big question facing our species – should we ever do what so many predict we will and end up colonising other planets or moons (see pp.71–3) – is how to overcome the issue of having sex in space. Given how long a trip to any planet outside our solar system would take, without procreation our species' time in space would certainly be short-lived.

We have a predicament: weightless sex isn't nearly as easy or as fun as it sounds. Zero gravity means it's quite literally tough to connect, and stay connected. And, thanks to Newton's Third Law, which states that 'for every action there is an equal and opposite reaction', a couple floating about in a gravity-less environment will have nothing to, um, push against . . . so their movements will kind of start to cancel each other out. It would be very tricky to err . . . push against your partner without them simply bouncing away from you. You can picture the rest yourself, but basically sex in space just doesn't really add up. There's also a lack of natural air currents to remove the sexy

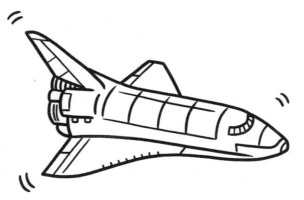

IF THE SPACE
SHUTTLE IS ROCKIN'
DON'T COME KNOCKIN'

sweat by usual means so things could get pretty slippery and uncomfortable to boot. Then there are the random floating objects hitting you while you're trying to do the deed, the lowered blood pressure (and we all know what that means), a lack of privacy and all sorts of other possible implications for conception and pregnancy. You get the picture. Tricky, right?

However, at the turn of the millennium a man named Pierre Kohler wrote a book exposing a series of experiments that NASA supposedly carried out secretly to work out if intergalactic hanky-panky was actually possible. According to Pierre, the study was a success and NASA established four positions that made sex possible in zero gravity using things like elastic bands and sleeping bag-like tubes. Articles were written, eyebrows were raised and lengthy denials issued by NASA. They called it a hoax, Kohler called it a genuine exposé and to this day no one really knows who to believe (hint: it's probably NASA).

CHIC FACT: In 2015 an IndieGoGo Crowdfunder was launched to help create the first-ever porno film in space. The team were hoping to raise millions, but instead they got just over £200,000 so the idea was shelved and a whole load of people's *Barbarella*-style fantasies were cruelly crushed.

While NASA's official standing is that sex in space has not happened yet, this hasn't stopped the rumour mill from churning full-speed ahead. Plenty of astronauts have had a questioning eyebrow raised in their direction since the first mission with both male and females launched in 1982. The most likely candidates according to the theorists are newlyweds Jan Davis and Mark Lee who went on a mission in 1992 soon after their unexpected marriage. Did they kick off the 500-mile-high club? Unsurprisingly, when asked, the couple refused to answer because, quite frankly, it's nobody's business but theirs.

Whether or not anyone has successfully had sex in space may be something that only a few people really know the answer to; after all, who truly knows what two astronauts get up to when their shuttle romantically disappears behind the Moon.

However, overcoming the issues faced when attempting space sex is one that does concern our entire species if we're serious about colonising a distant planet some day. It's a problem some people have taken seriously, among them the author and inventor Vanna Bonta, who developed a '2Suit' in 2006 to help navigate the difficulties of weightless nooky. The boiler suit-type outfits worn by the couple have large flaps at the front with Velcro strips on them. With Velcro space boiler suits, who needs lingerie?

Space isn't the only place where sex can be tricky; back here on planet Earth the seemingly simple act of getting down and dirty can be mired in problems. Luckily, as ever, it's science to the rescue.

Are Sound Waves the New Viagra?

Sometimes it's all too easy to lose sight of how simple sexual arousal really is. Sex is ultimately all in the plumbing. When a person becomes sexually aroused their brain sends electrochemical messages down their spinal cord to their genitalia. The blood vessels bringing blood *into* the genitalia *widen* to let more blood in. The blood vessels taking it *out* again *constrict* to prevent the blood from leaving. That's it! 'Vasocongestion' or, as we like to think of it, a blood traffic jam, within the sexual apparatus of both parties is all you really need, biologically speaking at least.

This simple localised increase in blood pressure is the engine of arousal. In men, this increase in blood pressure makes a relatively small, unobtrusive and easy-to-store floppy penis into a glorious rigid boner, handy for getting to those hard-to-reach places. In the female system the plumbing is obviously different, but uses surprisingly similar mechanism. The increase in blood pressure inflates the cylindrical walls of the vagina, increasing its circumference and depth, to ease access. Meanwhile, the stretching of the blood vessel walls serves another invaluable function – it allows the liquid component of the blood to accumulate on the inner surface of the vagina and produce a natural, water-based lubricant, making it nice and slippery down there to decrease friction during sex and prevent any damage. It isn't rocket science. It's evolution. It's a beautiful thing. Yet, for many, it can prove elusive.

For women whose genital vasocongestion isn't functioning well enough to make sex comfortable and enjoyable a little extra lubricant bought in a tube can help, but when a man's chap isn't playing dice simple remedies are few and far between.

For many with erectile dysfunction the arrival of Viagra truly revolutionised their sex lives. However, Viagra, which works by blocking an enzyme (see Geek Corner), is not without its drawbacks. Perfectly healthy men who sometimes try it out for recreational purposes can become increasingly reliant on the drug and eventually find themselves completely unable to perform without it. Side-effects from the drug vary from headaches and dizziness through to nasal congestion and sudden hearing loss. Wouldn't it be so much better to fix the problem without any of this pill-popping tomfoolery?

CHIC FACT: Many men who think they are impotent are in fact not. They may get erections in their sleep all the time, but simply don't realise it because they're unconscious when it happens. To find out one way or the other, the nocturnal penile tumescence stamp test was invented. This classic DIY test simply involves looping a strip of stamps around the base of the penis at bedtime. If, come morning, the perforations are broken then the problem must be psychological rather than biological. Worrying about sex can be one of the problems, creating a vicious cycle. Anxiety, for example, can disrupt the flow of messages from the brain down the spinal cord to the penis which keep the smooth muscles within relaxed and chock full of blood. You might want to bear this in mind next time you try to have sex in a public place.

GEEK CORNER: Impotence is caused by an enzyme that breaks down a substance called cyclic guanosine monophosphate, preventing it from doing it's job – namely causing the smooth muscles around every blood vessel in the corpus cavernosum (the sponge-like tissue in the penis) to release their grip. That lets it fill up with blood to produce and maintain an erection. Viagra (sildenafil) works by blocking this enzyme, allowing the cyclic guanosine monophosphate to fulfil its erotic duties.

Enter extracorporeal shock wave therapy (ESWT), which may sound like a music genre that can induce out-of-body experiences, but is actually a form of low-intensity sound-wave therapy that is applied five times a week, for several weeks, at six sites along the shaft of the penis. It can successfully raise boners from the dead – and if you are now picturing a crowd of willies swinging down the high street listening to Barry White on tiny headphones, we can only apologise.

In a trial of 112 men previously unable to have penetrative sex who underwent the experimental sound-wave treatment, a whopping 57 per cent overcame their sexual dysfunction without any chemical assistance. Interestingly, 9 per cent of those given placebo therapy had successful results too, which just goes to show how much sexual performance can be all in the mind. Best of all, ESWT is thought to work by stimulating the growth of new blood vessels within the penis (enabling sufficient turgidity for a sustained erection), which, if genuinely the case, means the benefits should last for some time.

CHIC FACT: Cardiovascular health is the greatest predictor of a healthy sex life, not surprisingly given that erections and vaginal arousal – the key prerequisites for comfortable penetrative heterosexual sex – are all about increasing blood in the genitals, which is, of course, pumped there by the heart.

Ultimately, this research could make the world a better place. After all, sex promotes bonding between humans. The more connected people feel to each other the greater levels of happiness can be achieved. And for every man who rediscovers his mojo there are usually two beneficiaries, both of whom can drift off to sleep together afterwards with a contented smile on their faces as they spoon in post-coital bliss.

Being able to have sex is one thing, but what about having babies? Some people face other issues in the quest to procreate, such as not having a womb. As ever, scientists are constantly striving to solve even seemingly unconquerable problems such as these.

Can You Have a Baby Without a Womb?

Anyone paying attention in school biology classes will know that one essential ingredient for having a baby is being in possession of a viable womb – a nice, safe, warm place in which a baby can develop and grow. Finding out you were born without a womb or having to have it removed prematurely for health reasons can be devastating for any woman who has her heart set on having children.

Malin Stenberg was one such woman, born without a womb and believing that her chances of one day having her own child were non-existent. In stepped her partner Claes Nilsson, a former pro golfer, who, naturally competitive by nature, refused to accept the status quo and was determined to find a way to fix this. Ultimately, Claes was right; after all, the genius scientific minds of the world would never let a small thing like a missing body part get in the way of the ambition to create new life. After much research the Swedish couple identified a pioneering project at Gothenburg University that was looking for ten women willing to try an experimental womb transplant. Each woman received a womb from an older sibling, relative or friend, who'd already had their children so had no further need for their child-building and bearing organs.

> **CHIC FACT:** In 1964 the first live births were recorded from a womb transplanted from one sheep to another.

In Malin's case it was an extremely generous friend, sixty-one-year-old Ewa Rosen, who decided to donate her womb. Incredibly, the ten-hour operation was a success, and the transplanted womb was not rejected. Sometime later, frozen embryos (created from Malin's egg and Claes's sperm) were implanted into the new womb and she successfully became pregnant. Nine months down the line and the world's first baby was born from a transplanted womb. They called their little miracle baby Vincent.

Since then, four more bundles of joy have been born following womb transplants, but, despite these successes, the procedure is far from simple. The first ever attempt in the USA only took place in 2016 and, due to some serious complications, sadly the womb had to be removed soon after the operation.

That said, overall, womb transplants are considered a huge medical breakthrough, bringing great hope to the 15,000 British women currently without wombs and countless more worldwide. At the moment this procedure depends upon finding a suitable living donor but doctors are now working on ways to grow wombs from scratch using DNA from deceased patients.

As for baby Vincent, he celebrated his first birthday in 2016 with his family and womb-donating godmother Ewa, who all got the happy ending they were hoping for.

So the wonderful world of medicine has transformed the lives of women who felt doomed to spend their life childless. Yet wombs are not the only reproductive organs scientists are looking to wave the magic wand of science over.

Grow Your Own Penis

Gentlemen – a question: have you ever had cock envy? Mathematically speaking, it's highly likely that you have. You will surely at some point have caught sight of another man's penis, when showering naked at the gym, sports hall or swimming pool changing room and ended up feeling, well, perhaps just a touch insubstantial. This is an experience that pretty much anyone with male parts has had at some point in life – unless, of course, you happen to be in the top fifth percentile. We all know that neither a man's length nor girth is the true measure of a man, yet it remains an issue that leads

to extensive bragging rights for some and a completely unnecessary kick in the self-esteem for the less well endowed. *It ain't the size of the boat* and all that.

Imagine a world where a smaller-sized man lusting after a larger member could have a surgical procedure to exchange his original member for a lab-grown replacement. No longer would the world have to endure male compensatory behaviours like driving flashy sports cars or superbikes at breakneck speeds along public highways.

> **CHIC FACT:** In 1993 John Wayne Bobbitt had his penis chopped off by his wife Lorena while he was asleep. His penis was subsequently surgically reattached and he went on to become a porn star all thanks to the publicity surrounding his bizarre story.

This procedure is not pie in the sky – a laboratory in Sweden has made it a reality. The science behind it is marvellous. First up, you need a donor penis; thanks to human mortality, these do become available fairly frequently. Next you use detergents to get rid of the cells from the donor penis. You're then left with a simple collagen structure that contains no tell-tale protein signatures to give the game away that the tissue originally came from another person's body, meaning no nasty organ rejection. Next up, the scaffold is seeded with the person's own smooth muscle cells, which control blood flow into the penis during erection. Then the endothelial cells are added to line the interior surface of the blood vessels. Once the transplant penis is fully rebuilt it has the form and shape of the original donor's penis, but it's made almost entirely out of the cells from the recipient's own body. So it's not quite as simple as 'Grow Your Own Baby Tomatoes' but you do get a new schlong with no threat of immune rejection.

> **CHIC FACT:** The world's first donor penis transplant took place in China in 2006. Sadly, the patient's immune system started rejecting the transplant shortly afterwards and he ended up requesting that it be removed just two weeks later.

Despite our earlier musings on cock envy, these incredible biomedical advances have not been driven by demand from rich, overprivileged Hollywood superstars in search of the ultimate plastic surgery job. Rather, it has all been inspired by a desire to help those with a genuine medical need. For men with genitalia deformed through disease or congenital abnormality, and for those born with genitals that don't fall cleanly into either the female or male anatomical mould, such transplants have the potential to be truly life-changing.

CHIC FACT: In South Africa the need for penis transplants is very high due to damage commonly caused by initiation-into-adulthood ceremonies and the like. In one recent case a twenty-one-year-old man had a transplanted donor penis surgically attached to replace the one he lost most of during a botched circumcision. He can now pass urine, get an erection and reach orgasm, but needs to take immunity suppressing drugs for the rest of his life, otherwise his body will reject the new penis as a foreign body. That's why these lab grown penises are such a breakthrough; no need for such drugs.

GEEK CORNER: Before surgeons got anywhere near humans with bio-engineered penises they had to prove that it worked on laboratory animals. In 2008 at Wake Forest Institute for Regenerative Medicine, North Carolina, twelve separate lab-grown penises were transplanted onto rabbits. All tried to mate, eight managed to ejaculate and four of these bunny encounters resulted in offspring.

Thought sex between humans was tricky, strange, exotic, freaky, awkward, exciting, bizarre and mired in problems? Wait 'til you hear about the weird stuff animals get up to.

Weird Sex in Nature

If you've ever watched one of the countless documentaries about the car-humping, horse-loving, orgy-partaking fetishes of the human race you'd be forgiven for thinking that when it comes to sex us humans are pretty odd creatures. However, spend a few moments reading about the bizarre sexual habits of the rest of the animal kingdom and you may start to think our tastes are comparatively bland.

Mandarinfish

Male mandarinfish can, quite simply, be plain rude. They wait until a loved-up couple are ready to get down to business and then pounce on them, uninvited. You see, when the poor, unsuspecting, mating pair are ready to do the deed, the female releases a cloud of around two hundred eggs so that her mate can release a whole load of sperm over it. Job done, right? Not if another male is lurking anywhere nearby. He will lie patiently in wait for the right moment, swim in at the last minute, metaphorically unbuckle his tiny trench coat* and then spray his own sperm all over the female's cloud of eggs. What a creep! The aim is to gazump the original suitor's sperm with his own and that way maybe a portion of the eggs will be fertilised by him instead. Presumably he missed the memo when we were told 'three's a crowd'.

Flatworms

Flatworms really take the biscuit when it comes to weird and, quite frankly, offensive sex. Not only do they have two-headed dagger-like penises, which sounds scary enough, but they also have penis fights with other flatworms to establish which one of them gets to be the father. It's a Jerry Springer episode just waiting to happen, with dagger-headed penis fights instead of throwing chairs.

You see, all flatworms are hermaphrodites – with both male and female parts. However, the problem is that, when it comes to mating, neither of them wants to be 'the girl' (the sexist so-and-sos). To make that decision, the two macho flatworms whip out their dagger cocks – which, by the way, they also use to hunt prey – and begin a penis sword fight, which can last up to an hour. An *hour*. During this wormy cockfight each is effectively trying to pierce the other's skin with their genital dagger to inject them with sperm – there's definitely a flatworm *Pirates of the Caribbean* spin-off idea forming here.

The duel ends when one hits the mark, but instead of the swash-buckling winner running off with the fair maiden, the champion in this case gets to ... impregnate the loser. The victorious 'father' worm buggers off – no doubt to do battle with more hermaphrodite worms

---

* We may be taking anthropomorphism to a whole new unacceptable level here but fish in tiny little outfits is definitely something we could support.

and spread its seed yet further – while the loser is burdened with the pregnancy and having to play the role of 'mother'.

Argentine Lake Ducks

If you ever glanced at a male Argentine blue-billed lake duck and thought he looked pretty pleased with himself, then it may not surprise you to know that these ducks are the Dirk Digglers* of the animal kingdom. Not only do they hold the Guinness World Record for the largest penis of any bird, they also have the longest penis in relation to their size for any vertebrate. The average penis size for the plucky duck is a tear-inducing 17 inches which, to put that in context, is 3.5 inches larger than the current known record holder among humans. But size isn't all their lady counterparts need to worry about – it's been suggested that some male ducks use their incredibly long corkscrew-shaped penises to literally lasso females and drag them over towards them.

Anglerfish

Worried that the mummy flatworms and lady Argentine lake ducks were getting a raw deal? Well, nature has a way of balancing these things out. The poor old male anglerfish is considered pretty worthless from the moment it is born. Small, weak and unable to gather much food, he is destined to spend most of his life hungry.

However, the male anglerfish has a brilliant sense of smell and can skilfully sniff out the pheromones of a female, who happens to be far bigger, far fiercer and far more adept at catching prey. She even has a glowing spine on top of her head, which lures in the unsuspecting prey (other fish, not the male of her species) so that, at just the right moment, she can snap her jaws to chomp down on the tasty morsel.

The male fish, seeing his fearsome new warrior lover as the answer to all his prayers, instinctively goes up to her and gives her a little nibble. As he does so, he releases an enzyme that quite literally dissolves his own lips and melds them onto our Xena the

---

* If you're wondering who Dirk Diggler is you need to go and watch *Boogie Nights* right now! The brilliant 1997 hit film directed by Paul Thomas Anderson tells the story of an extremely well-endowed male porn star named Dirk Diggler, a character Anderson first imagined in his 1988 mockumentary *The Dirk Diggler Story*.

Warrior Anglerfish's body. Shortly afterwards his face and body start melting too, allowing him to fuse completely with her body, so that she can never, ever, ever, *ever* get rid of him – and you thought you'd seen it all when it came to clingy boyfriends? The male fish gets to feed off her nutrient-rich diet and circulatory system for evermore like a weedy little succubus. Meanwhile, the female – who gets to fuse with *multiple* males (saucy lady!) – gets all the sperm she needs for reproduction in return, and to swim about with her newly fused male harem, like the fierce little temptress that she is.

Hippos

There's nothing to get you in the mood quite like the fragrant smell of a hot, stinking poo. At least, not if you're a female hippo. The male hippo will wait until he's caught the attention of a potential lady hippo mate and then take a gigantic dump to lure her in. Mmmm . . . Just in case that hasn't quite seduced her, he whips his tail around, wafting the juicy smell over in her direction so she has no chance of missing it. It's certainly a cheaper alternative to CK1.

Bonobo Chimps

Our primate cousins could teach us a thing or two about the expression 'Make Love, Not War'. Instead of settling disputes over land or food with a traditional fight, the peaceful bonobos choose the rather more pleasant option of sexing it out. And that's not all. These sex-loving chimps don't only use sex to settle contentious issues, but for bartering, reconciliation, pleasure or even just to say hello. It's like a bonobo-tastic Woodstock, taking free love to a whole new level. They're not too picky either, happy to make whoopee with any sex, any age, anyone, except their mothers – even for the horny bonobos incest is a step too far. As for how they like to do it, they're more like us than any other creature, enjoying French kissing, oral sex, mutual masturbation, same-sex intercourse, genital rubbing and face-to-face sex. The Bonobo *Karma Sutra* could teach us a thing of two . . .

GEEK CORNER: Bonobos have been observed in the wild by scientists for hundreds of years. Yet their hypersexual activities so offended the morals of Victorian anthropologists that they pretended it wasn't happening. Even then, the similarity between bonobo chimps and humans was self-evident (we've since discovered that we share over 98 per cent of DNA, indicating a common ancestor a mere couple of hundred thousand years ago). The Victorians, however, completely neglected to report any detail of the orgiastic behaviours they must have observed – bonobos will typically rub genitals with another individual once every hour or two – so the suspicion is that it simply flew too squarely in the face of the prevailing religious beliefs.

Dolphins

Along with bonobos, dolphins are one of the few creatures apart from humans to have sex purely for pleasure, not just when reproduction is possible. They too are partial to a bit of same-sex intercourse. However, one researcher accidentally found out that dolphins like to partake in some very surprising behaviour.

A mirror was set up in a dolphin pool for one of many studies investigating which species of animal have sufficient brain power to recognise their own image – a classic measure of self-awareness. In humans, we start to cotton on to the idea that the person looking back from our reflection is us at around eighteen to twenty-four months old. Some great apes, magpies, elephants and orcas have passed the so-called 'mirror test', too, and dolphins are now known to be particularly adept at it.

In this particular test, dolphins that were marked on their body with black marker pen would swim over to a mirror in the tank to check themselves out. Dolphins marked with a 'sham-mark' (a temporary mark on their body made with a water-based pen that immediately washed off) would also go over to the mirror and continue to flip about searching for the mark on their body for some time. Clearly the dolphins were aware they were looking at themselves in

the mirror, not that unexpected given what we know about their enormous capacity for thought and reasoning.

However – and here's where it gets saucy – during the course of the research one scientist noticed that, in between testing, the two male dolphins, named Delphi and Pan, had placed themselves in front of the large mirror and were sticking out their penises, then proceeding to have full sex with each other in front of the mirror. Wondering if the choice of location for their lusty antics was just a coincidence, the researchers covered the mirror and the dolphins promptly went back to shagging elsewhere. However, the moment the mirror was uncovered again the dolphins went back to doing it in front of the mirror. The researchers were convinced it could mean only one thing: the dolphins were *enjoying* watching themselves having sex. Who knows, maybe one day they'll discover they like rotating water beds and leopard print flipper-cuffs, too.

Drone Honey Bees

In order to reproduce, swarms of drones compete, mid-flight, in an orgiastic frenzy to penetrate a virgin queen once she's been, quite literally, sniffed out by her suitors. The sexy kiss chase itself lasts up to thirty minutes, while the penetration part lasts a matter of seconds (a scenario we very much hope doesn't sound too familiar to you). However, that's only the start of the back-flipping, testes-exploding, suicide-making, orgiastic, sexual habits of the drone honeybee. More on that coming next.

That's just a few of the more bizarre sexual habits of our animal friends. We're pretty sure that when The Bloodhound Gang sang: 'Let's do it like they do on The Discovery Channel' that wasn't exactly what they meant.

Not only are the sexual habits of the birds and the bees extraordinary to beehold (boom boom) but bee semen even seems to have some medicinal qualities. And, as you're about to find out, it might turn out to be key to the very survival of their species.

Power of Bee Semen

For some time now we've known that bees are pretty damn important. Not only do they make delicious honey and impress us with their ability to communicate through the medium of dance (Michael Flatley fans, you ain't seen nothing yet!), but they are also partly responsible for the not-so-small feat of keeping *us* alive.

How so? Next time you sit down for a tasty meal have a good look at the food on your plate and take a moment to thank our little honey-making friends. Around a third of everything you see there will be somehow reliant on the continued existence of bees. They either directly pollinate the yummy fruit and veg, or they pollinate the crops that feed the livestock which provide us with our meat and dairy.

Pollination – for those of you desperately casting your minds back to biology GCSE classes when you were too busy passing notes to pay attention to teacher – is an essential part of reproduction for many flowering plants. Honey bees buzz about, landing on different plants and flowers to collect nectar and pollen to feed themselves and their young, and in doing so catch a whole load of pollen on their furry little bodies. Some of the pollen naturally drops off when they visit other plants and so pollen grains from the male parts of one plant (the anther) are transferred and dropped onto the female part (the stigma) of another plant. Fertilisation occurs and wham, bam, thank you, bees – all sorts of tasty fruit, vegetables and livestock feed can blossom and grow.

Bees may not be the sole pollinators – wind, birds and other insects and animals all play a part – but honey bees *are* hugely important, with more than seventy varieties of crops in the UK alone highly reliant on them. Enjoy apples, apricots, blueberries, cherries, cucumbers, pumpkins, pears, blackberries, plums and squash? Well, you may have to kiss goodbye to all those, and many other delicious foods besides, if bees aren't around to carry out their pollinating duties. There's also, as mentioned, the vital role they play in production of plants for livestock feed, which in turn impacts on meat and dairy production. Still don't fully appreciate how important bees are? Try adding in that they make around six thousand tonnes of honey in the UK every year and are estimated to contribute around £400 million to our economy. Even the staunchest of capitalists ought to be swayed by that.

We're all on the same page now, right? Bees are vital to the survival of flowers, fruit, veg, livestock, honey, us and our economy. So, what's going wrong with them? And what's it got to do with sperm?

The Threat to Bees

You may have noticed that bees have been in the news quite a lot in the last decade, and not because their media spokesperson is advocating the joys of honey. Pesticides, climate change, bad bee-keeping practices, habitat loss and disease have all contributed to serious declines in bee populations. There's also something known as 'Colony Collapse Disorder,' where the majority of worker bees in a hive suddenly disappear or, more likely, die, with bee-keepers typically reporting losses of up to 90 per cent of all their bees. One insecticide in particular has been taking a lot of the flack. Neonicotinoids – which account for a third of all insecticides globally – have demonstrated some pretty heinous effects on bees as well as other insects, wildlife, plants and soils. A partial EU ban on them has helped somewhat, but the challenge faced by the world's bees is nowhere near over.

As if all this bee-misery wasn't bad enough, enter the parasitic mite varroa, which is spreading a highly contagious and deadly disease among bees. The mite bites the bee, injecting it with a deadly 'deformed-wing virus' (symptoms include deformed wings, legs and bodies and sufferers have a life span of just forty-eight hours) which can destroy an entire honey bee colony. Meanwhile, another fungal pathogen known as *Nosema apis* – which can be transferred from bee to bee through bodily contact, faecal matter or sexy times – is busy causing nosemosis, the most widespread of all adult honey bee diseases. The pathogens work by invading the cells of the digestive system, specifically those lining the bees' guts, quickly multiplying and filling them up with spores, which are excreted by the bee later on. If the bee colony is in a good state most of the bees will do their business outside the hive, but in times of stress caused by, for example, bad weather or lack of food, the poor sickly bee has to defecate in the hive, which means the disease quickly spreads, potentially causing total colony collapse.

So far, so doomworthy, but is there *anything* we can do to save our black and yellow stripy friends?

Bee Spunk to the Rescue

It's the kind of thing we can imagine men around the globe being absolutely delighted by: semen may be the wonder solution. A bunch of Australian scientists got together to try and figure out how to tackle the problem of *Nosema apis* and started to test a whole load of different biological defences. We're not sure which bright spark decided to test the bees' love juices, but whoever it was struck icky-sticky gold, because it turns out it was the answer to all their fungal pathogenic nightmares. Apparently bee seminal fluid has some powerful anti-microbial qualities, which destroy the fungal spores in not one, but two different ways. The protein part of the semen causes the fungal spore to germinate early and die off, while the non-protein part simply kills off spores directly.

If you're wondering how the hell they managed to make a bunch of bees ejaculate so that they could test the powers of their semen, you're not alone. And, no, they didn't put them in tiny booths with mini bee porn and watch them exclaim: 'Ooh, look at the gorgeous hairy bum on that one!' They did something possibly even stranger. They put two hundred drones – whose collective purpose in life is to make sweet love to the queen bee, serve her every need and tend to the resulting baby bees – into a cage and gassed them with chloroform, which induced ejaculation (hey, who are we to judge?). The semen was then collected in a tiny pipette before being rubbed onto fungal spores collected from an infected hive. Lo and behold, within five minutes most of the spores had been successfully killed off.

The only problem? For drone bees, ejaculation is fatal. Yup, these love-making little horn-dogs ejaculate so powerfully that afterwards it's game over for them. Back in the wild, when drones get down to business they eject their semen with such tremendous speed and force into the queen's lady parts that a 'pop' noise can be heard, audible even to the human ear. The drone becomes paralysed and does a backflip, his testes explode and his penis comes clean off, left behind in the queen's vagina. Death by orgasm, what a way to go.

However deadly to the drone this delivery may be, bee semen is vital. Not only for continuing to ensure bee reproduction takes place and new generations of bees are born, but now, more vitally than ever, to prevent the spread of the deadly fungal pathogen *Nosema apis*. Bee semen doesn't just help bees to thrive, it may also be the key to protecting our own food supply and – with ever-growing human populations – our ultimate survival.

Huey Lewis had it all wrong when he sang 'The Power Of Love'; the power of bee semen is far more impressive.

Final Thoughts

There you have it. Wasn't so bad after all, was it, talking about all the strange, ridiculous, dirty, sexy, tricky, fun stuff we and our fellow critters like to do when we're naked. Whether we end up getting freaky in space, using sound waves as the new Viagra or growing our own penises, sex is here to stay. It has to be. After all, without it we would simply cease to exist . . . and speaking of ceasing to exist, it's time now to turn to the next doom-worthy chapter.

7

It's the End of the World As We Know It

There are more than seven billion people on the planet today and the majority of us, in the developed world at least, are taking full advantage of the incredible advances in technology, innovation and science that add to our convenience and quality of life. However, such benefits come with a price tag (and we're not just talking economics!). If you tot up the overall impact of human advancement, the cost may well be the demise of our very own home, planet Earth.

In this next chapter, we'll be careful, of course, to avoid the kind of 'end of the world is nigh!' sign-toting ranter. The termination of life on Earth has little to do with how carefully you studied the ancient instructions in whatever religious handbook happens to be popular in your community. Instead, it has *everything* to do with the human tendency to focus entirely on what we want now, rather than considering the consequences of our choices in the future, coupled with threats from major cosmic events over which we have little (if any) power.

Most of us realise that praying isn't going to make a difference, but using science to figure out what's going on and inventing some new technology to bung the hole in the hull of the Good Ship Planet Earth might just keep us afloat for a few more centuries. In reality, the superheroes who might fly in to save the day have no special powers beyond a gift for science. They won't wear tight luminescent body stockings and capes (not at work, anyway) but instead don lab coats and safety glasses to take careful measurements of our oceans, atmosphere and the Sun, to design new power-extraction techniques and invent ingenious new ways to protect endangered species.

It's time to steel yourself, people, to pull your heads out of the sand and take stock of what's in store for our particular lump of planetary rock over the next thirty years or so. But, first, let's take a little wander down Earth's apocalyptic memory lane.

CHIC FACT: A large-ish meteor – heavier than the Eiffel Tower – came into the Earth's atmosphere over northern Russia on 15 February 2013. It illuminated the morning sky over the town of Chelyabinsk with a blinding light more powerful than the Sun. The whole thing was caught on the dashboard-cams of several drivers as they made their way to work in the morning commute. Using those videos along with data from seismic detectors across the world it was possible to calculate the size (20 metres), speed of entry (40,000 mph) and explosive power (equivalent to 500 kilotonnes of TNT or twenty-five times that of the Hiroshima atomic bomb) of the only superbolide meteor we are ever likely to see in our lifetimes. Luckily for us, in the context of a human life span they are pretty damn rare!

The First Five Apocalypses

If we asked you how many apocalypses our planet had already experienced you'd probably close your eyes, think of the poor old dinosaurs and say: 'one, maybe two' – unless of course your brain was actually fully engaged when you read the above, in which case you'd confidently tell us 'five'. Smart arse.

And you'd be right. Over the course of the four and half or so billion years the Earth has been around, there have been five full-on apocalypses that we know of – mass extinctions that permanently killed off entire species in vast swathes – each commonly attributed to what insurance companies like to refer to as an 'act of God'.

So here's a handy little recap of mass extinctions gone by – if this book was a film this bit would definitely be played out as a montage with some dramatic orchestral background music:

1. The first mass extinction was around 440 million years ago, killing off 60 per cent of life in the sea. It's likely that this one came about as a result of global cooling, which caused the formation of huge glaciers and a massive drop in sea level as a direct result. Poor old fishies and ancient sea monsters of the deep.

CHIC FACT: Life took between ten and thirty million years to recover from past mass extinctions, i.e. at least forty times longer than we humans have walked the Earth.

---

2. The cause of the second, 360 million years ago, is a bit more of a mystery but, whatever its origin, the sea took a massive hit once again and 70 per cent of all marine species were completely wiped out. Sucks to be a sea creature!

3. Then, around 251 million years ago, the third and worst ever mass extinction event annihilated 96 per cent of *all* species. The remaining 4 per cent eventually evolved into all the life we now see on Earth (imagine what the species of the world might have looked like today if this had never happened!) The most likely cause of this one was a bunch of super-volcanoes in Siberia pumping out immense quantities of lava onto the surface of the planet accompanied by a gargantuan release of methane that poisoned land, sea and air.

4. The fourth took place somewhere between 199 and 241 million years ago thanks, it seems, to yet another flood of lava, this time erupting from what is now the central Atlantic. This was the event responsible for splitting up the supercontinent of Pangea and wiping out half of all sea life

in the process. Those fish really can't catch a break, can they?

5. In our lifetimes, the event that wiped out all the big dinosaurs around sixty-five million years ago in our planet's fifth mass extinction has gone from being a huge mystery* to a near certainty. From time to time our planet carelessly strays into the path of one of the millions of rocks that, like our beloved Earth, have been spinning round and round the Sun for billions of years, boasting girths that vary from a measly 10 metres right up to the frighteningly vast 10 kilometres. The Chicxulub Crater, which is buried underneath the north-western shoulder of Mexico's Yucatán peninsula, is the result of the impact of an absolutely ginormous meteor. We reserve the use of the word 'ginormous' exclusively for truly, extraordinarily, immensely large things, like a six-mile-wide flying mountain crashing into the Earth with an explosive force equivalent to 240,000 megatonnes of TNT† – creating a 190-mile-wide‡ crater stretching twelve miles deep under the Earth's crust.

The impact of the meteor would have kicked up untold amounts of rocks, dust and fireballs – blocking out the sun for around a month (halting the growth of plants/food resources) and choking the air with dust. Now it may just be a coincidence that this immense celestial body touched down sixty-five million years ago (a period known to geology as the Cretaceous-Paleogene boundary) precisely the point in time after which the fossil record becomes suspiciously devoid of T-Rex and Co.'s bones. But then again, Tupac could also be alive and well, sitting on some tropical island with Elvis, JFK and the Easter Bunny, planning their comeback as the next supergroup.

---

* There have been over 100 different extinction theories proposed in the past 40 years, including poisonous plants, volcanoes and dinosaur mass suicides!

† By contrast, even the largest man-made bomb was five million times LESS powerful than this.

‡ The initial evidence suggested the impact zone created a ring 110 miles across but more recent evidence indicates that this is merely an inner wall!

GEEK CORNER: Father and son duo Luis and Walter Alvarez were the first to propose, in 1980, the theory that an asteroid impact was what killed off the dinosaurs on the basis of the discovery of a layer of iridium-enriched clay. Iridium is relatively rare on Earth but fairly common in space so it seemed logical that something from space must have exploded on impact with our beloved home rock, scattering iridium far and wide. At that time, however, the Chicxulub crater had yet to be discovered.

CHIC FACT: Thought the only use for wishbones was to make a wish after a roast chicken dinner? They also provide a key link between birds and dinosaurs, with both sharing this anatomical structure known formally as a furcular. In fact, according to Mark A. Norell, Chairman of Palaeontology at the American Museum of Natural History, New York: 'Dinosaurs are not extinct, we just call them birds.'

> **GEEK CORNER:** Investigation of the Chicxulub crater has slowly ramped up since its discovery twenty years ago when potholers realised that a series of huge sink holes were arranged in a perfect arc, indicating that they might actually mark the perimeter of a super-massive asteroid impact. The latest thinking is that the high-velocity impact of such a large mass causes the Earth's crust to behave a bit like a liquid. The animations showing what happens bear a striking resemblance to super-slow-motion footage of a water droplet bouncing back up after striking the surface of a body of water.

There you have it, a handy little guide to the times-species-have-been-wiped-out-en-masse. As for the next one, it may be just around the corner.

Earth's Sixth Mass Extinction

Look back over millions of years' worth of fossil records and you'll see that exotic new creatures pop up all the time. After a few thousand years, or millions if they're lucky, some of these species then disappear off the face of the planet never to be seen again – just like the dinosaurs. New creatures constantly evolve while existing species die off completely. Such are the rhythms of life on Earth, the ecological dance between biology and geography.

As we saw earlier, every now and then something really big happens that wipes out many species with breathtaking efficiency. Recently it's become apparent that, for the first time since the dinosaurs were choked to death by the endless plumes of dust and fumes

kicked up into the atmosphere by that humongous meteor, species extinction once again is outpacing species creation.

Scary stuff. And this time it's different. For once, super-volcanos, glacier formations and massive meteor impacts aren't to blame. This time the cause is a phenomenon that has only been around for a measly 200,000 years, a mere blink of an eye in the long history of our planet. The cause of all this lethal destruction to so many species is . . . us. Pesky humans.

GEEK CORNER: There are around 1.9 million species of animal on our planet and around 450,000 species of plant. Their rate of extinction is 1,000 times more than it would be if humans didn't exist. Currently 100–1,000 of every million species on Earth die out each year. Before we humans came along the rate was less than 1 per million species annually.

Ever heard of the Four Horsemen of the Apocalypse? The dastardly beings were first mentioned in the New Testament's Book of Revelation. Named as War, Famine, Pestilence and Death, they were predicted to ride the Earth like wild creatures at the end of days. However, in modern times things have changed. So great is the destructive nature of our own species that the new 'four horsemen' of the sixth mass 'apocalypse' are the various ways we humans are interfering with the natural order of the planet to cause the extermination of countless species. They are:

- *Globe-trotting.* Thousands of plant and animal species have been ferried to and from every corner of the globe, ending up living in places they'd never have reached without human intervention. When you think of a typical farmyard, pigs and chickens will probably be among the first animals to spring to mind. For us, at least, they are synonymous with the English countryside, yet are actually indigenous to Asia. These particular immigrants to the Western world co-habit with the locals quite contentedly and most of us have no idea they originated so far away.

However, it's not always happy families. These days, lots of extinctions have been caused by introducing (intentionally or otherwise) species from foreign lands into new continents, only for them to outcompete local species for resources, or become pests and predators against which the locals have no naturally evolved defences. Ever heard of the dreaded cane toad? Native to South America, they were introduced to Australia in the 1930s in the hope that they would devour the beetles that were destroying the valuable sugar cane crops. Instead, they multiplied like rabbits, ate almost anything that came into sight, and wreaked all sorts of havoc on the local populations of reptiles, birds and other critters who died after trying to eat the poisonous toad, or couldn't compete with it for food and shelter. Cane toads are now an uncontrollable pest, with a devastating effect on Australian biodiversity. However, the fat little beasties aren't all doom and gloom – they also entertain Aussie locals who bet on them in the popular cane toad races.

- *Resource Monopoly.* The second horseman has nothing to do with hotels on Park Lane and a 'Go to Jail' card, and everything to do with humans monopolising our planet and its limited resources. We are the number one, top-dog predator on both land and sea, greedily sucking up biomass* in the form of animals and vegetation like nobody's business. We hog a whopping 25–40 per cent of the planet's net primary production (which includes land and biomass) for our own use, and we hoover up vast amounts of fossil fuels from the ground to use for energy production. All this pillages the Earth of its natural resources and disrupts the natural environment for other species, diverting food, living space and water away from them.
- *DNA-meddling.* The third horseman is human intervention in the evolution of other species. We divert the trajectory of

---

* Biomass is organic matter from living (or once living) organisms and includes vegetation, animals, etc.

the natural process of the 'survival of the fittest' to suit our own needs. This includes anything from selective breeding of animals and plants to improve disease-resistance or adapt them for domestic use to the relatively modern invention of direct genetic modification by science.

- *The technosphere.* The fourth horseman of the human-induced apocalypse is the extra demand placed on the planet by the 'technosphere' – a term coined by Peter Haff – which includes all of the resources hogged for the development and maintenance of various technologies and human interactions with it. Haff argues that technology is transforming the way our planet would naturally function in such a big way that it will lead to mass extinctions.

CHIC FACT: There are fewer than 1,600 pandas left in the wild and millions are spent trying to stop them becoming extinct. BBC *Springwatch* presenter and naturalist Chris Packham famously said that we should simply 'let them go with a degree of dignity', pouring the cash into the preservation of other species instead. But Chris ... they're SO DAMN CUTE!

Together these four horsemen are running rampant across the planet, gobbling up resources, destroying natural ecosystems and driving more and more species to extinction every single year. But the situation isn't completely hopeless ... yet. If we can fundamentally rethink our relationship with the planet and acknowledge that it is not only in our collective best interests, but also easily within our power, to change how we operate, then this dire situation can be salvaged.

We could learn a lot from the various indigenous tribes across the world who have acted as respectful and responsible guardians of their local environment for millennia. If we can start to see ourselves as guardians of the planet, rather than acting like schoolkids 'running riot through a sweetshop' (as one *Guardian* writer described it), as we have done for the past few centuries, we may still have one last chance to stop the rot before it really is too late.

CHIC FACT: The Costa Rican golden toad has been wiped off the face of the planet by the chytrid fungus. This ancient fungal organism seemed to co-exist quite happily with amphibians for millions of years in Africa and North America, but once it was transported to the rainforests of Central America – probably accidentally by clueless-but-well-intentioned backpackers – it was soon game over.

---

*Now we've identified that man is effectively the scourge of the Earth (*quelle surprise!*), let's take a closer look at some of the ways in which we are rapidly destroying our planet. Sounds like a lovely way to spend an afternoon, doesn't it? Here goes . . .*

Killer Face Scrubs

Ahh – the good old morning routine: You get up, brush your teeth, comb your hair, wash your face and then unknowingly poison marine life and mess up the entire ecological balance of our planet. Sound familiar? Possibly not. But the reality is that the simple act of exfoliation may be causing untold damage to our seas.

The tiny microbeads used to remove dead skin cells from our faces during our morning scrub are more often than not made of tiny pieces of plastic, less than five millimetres in diameter, collectively referred to as 'microplastics'. Once these get washed down our sinks they get into our sewerage systems, then our rivers and eventually our seas and oceans. This forms a not-so-delicious 'plastic soup', which fish and other sea beasties appear to be slurping down with wild abandon.

CHIC FACT: A recent study looking at a sample of 504 fish caught in the English Channel, including ten different species, found that a third of them had the offending microplastics lodged in their bellies.

If you're unfamiliar with the molecular qualities of microplastics you'd be forgiven for not feeling overly alarmed by the rapidly increasing plastic content of our oceans. However, not only are these non-biodegradable microplastics indigestible to the marine life that are merrily chomping them down, but the beads themselves actually act as tiny little sponges, absorbing all sorts of toxins, chemicals and nasty organic pollutants from the water in which they swim. All this eventually turns them into tiny little bullets of poison.

And guess where the sea creatures that chow down on these tiny toxic balls end up? That's right – on our plates and in our bellies! Fish and microChips anyone?

CHIC FACT: We produce 300 million tonnes of plastic per year, equivalent to the total mass of the entire human race.

Add to this the devastating effects of the larger plastics accumulating in giant mid-ocean rafts, scattering across remote beaches and poisoning marine plant life, and it's pretty clear that something needs to be done. And it's not only face scrubs that are the culprits, either. Toothpastes, body scrubs and a whole host of other beauty products contain these damaging beads.

We need to fix this, and fast, but what can we do? Well, first and foremost, the big fat-cat cosmetics companies need to stop using plastic microbeads in their products and we *all* need to stop buying the products that contain them – simple as that! But how can you possibly tell which products do and which don't? Luckily for us someone else has done all the hard work: two Dutch NGOs (the Plastic Soup Foundation and the North Sea Foundation) developed a smartphone app called Beat the Microbead, which lets you scan a beauty product and tells you if it's an offender or not. There are plenty of plastic-free options available.

So, the next time you're reaching for a facial scrub, check it out first, and if it does contain plastic microbeads ask yourself this: is a smooth skin from that particular product really worth poisoning yourself and the rest of the planet?

CHIC FACT: America has woken up to the dangers of face scrubs! In December 2015 Barack Obama banned the use of plastic microbeads in cosmetics products. The new law is called the 'Microbead-Free Waters Act of 2015'.

While a huge amount of marine life is being threatened by our plastic-happy morning facial routines there's one sea creature that is still going strong; in fact so strong that it's starting to cause some serious problems.

The Jellyfish Apocalypse

Jellyfish are right little buggers, aren't they? They swim around waiting for the perfect moment to scupper all the peace and tranquillity of your holiday, by wrapping their toxic tendrils around your legs and stinging all the joy out of you. They get themselves tangled up in boat propellers and other marine machines. The boxfish variety have killed over 76 people since 1884. And we hear they cheat at poker, too – slippery little bastards.

OK, so that last one may have been a lie, but there is one more to add to the list – they are threatening us with a 'jellyfish apocalypse'. And no, that's not a new Japanese TV series, it's a real threat to our seas, our species and our planet.

The big issue is that there are just too many jellyfish and their populations are growing at an exponential rate with potentially catastrophic results. Jellyfish smacks* are growing to such immense sizes that they've even started clogging up the water pipes of nuclear power

---

* A group of jellyfish is known as a smack, a flutter or a brood.

plants, causing emergency shutdowns (although better a shutdown than a meltdown).

The main reason? Us humans; surprise, surprise. We've been busy screwing up the oceans with fertiliser run-off, toxic oil spills and all plastics great and small, making life very hard for all the poor old sea creatures ... all, that is, except for jellyfish, which have been in our seas for more than 550 million years and are going stronger than ever. Our blobby little nemeses are hardy little characters, you see, multiplying with wild abandon in circumstances where other fishy creatures struggle to cope and surviving even toxic conditions.

CHIC FACT: In the summer of 2011, surfers flocked to the Florida coast to take advantage of the big swell produced by a handful of offshore hurricanes, only to find themselves confronted by a shoreline brimming with moon jellyfish – some the size of bicycle wheels – which triggered the shutdown of a nearby nuclear power station. Earlier that year giant jellyfish blooms also caused the shutdown of power plants in Japan, Israel and Scotland.

Even if we put pollution to one side, global warming itself strips oxygen from the water. Together with overfishing, this knocks out other marine species that would act as competition, leaving more tasty food for the jellyfish to thrive on. If that wasn't enough of an advantage for our slimy friends, they are also thought to have the capacity to promote global warming themselves! All the poo and goo that these massive flutters of jellyfish release into the ocean as a result of their own feeding behaviour provide food for sea-borne bacteria

that in turn release carbon dioxide into the atmosphere to further accelerate the greenhouse effect.

You may think you've already heard enough to resign yourself to a world where jellies are the overlords of the sea, but, sadly, we're not done yet. They also have a phenomenal capacity for multiplication – large broods can produce more than a million eggs, each capable of producing twenty to forty new jellyfish offspring, and they can even multiply all by themselves. And, like true super-villains, some are even immortal!

GEEK CORNER: Several jellyfish, such as the moon jellyfish, reproduce both sexually and asexually. The sexual part results in larvae that drop to the ocean floor and turn into polyps. These polyps are where asexual reproduction comes into the equation because every polyp can bud to produce twenty-plus adult jellyfish. In hard times they can even form a tough cyst, sit tight until circumstances improve, at which point they can trigger a bloom to start the reproduction cycle all over again. Jellyfish blooms used to happen only occasionally. Now they occur annually.

It's safe to say that a jellyfish takeover is well and truly within our sights. As well as the aforementioned global warming, clogging up nuclear power plant pipes, wiping out other sea-life and stinging the bejesus out of us; worst of all, the looming jellyfish apocalypse will also act as a conveniently visible benchmark for the overall health of the planet, with every sting a karmic reply to our flagrant disregard for the best interests of our environment.

It's unlikely that the world will wake up in time to prevent the jellyfish from taking over our seas. Sadly, the world is still chock full of global warming denial – most people seem entirely unconcerned by what's happening to our planet beyond their own front yards. But we reckon that if their beach holidays were routinely ruined by seas choked with slimy, stingy jellyfish, then even they'd be forced to sit up, take the declining health of the planet seriously and finally stop hiding their heads in the proverbial sand.

CHIC FACT: The Japanese Nomura's jellyfish are the sumos of the sea, weighing a whopping 200 kilograms. On the other hand, the five-millimetre Irukandji jellyfish, indigenous to Australia, may be small but it packs one of the most painful stings known to man. While most jellyfish have their stingers exclusively on their tentacles, this nasty little piece of work also has them all over its belly and, worst still, they can be fired through the water.

Jellyfish may sound like an unusual threat to our very existence but we reckon this next one is even more unexpected. According to Sir David Attenborough himself, misogyny may be killing the planet.

Is Misogyny Killing the Earth?

We have a confession to make. When Sir David Attenborough walks into a room we start to fangirl. Forget Directioners or Beliebers, we're bona fide Attenbabies.

This also means that when Sir David says something we damn well pay close attention. So, when we attended a talk by the great man himself and he declared that equal rights for women was the key to saving our planet, we made sure to take note. But what did he mean by this? After all, anyone with half a brain knows that equal rights for all is a positive thing, but how could this possibly translate into saving the world?

CHIC FACT: Women got the vote in the UK in 1918, the USA in 1920, China in 1949, Switzerland in 1971, South Africa in 1994 and Saudi Arabia in 2015.

The answer is deceptively simple. The single greatest threat to the survival of planet Earth – according to Sir David – is the rapidly increasing human population. Makes sense, right? After all, we've already seen some of the many ways our species is screwing things up. Today there are approximately 7.4 billion people on Earth and unless there's a dramatic change in our collective behaviours this will reach around 9.7 billion by 2050. That would take us to just a smidge under our planet's 'carrying capacity' – the maximum amount of humans the world is considered capable of supporting. Many think that at around ten billion it'll be game over, giving us just over thirty years before apocalypse time. Great!

And if you're wondering what all this has to do with misogyny, then here's another Sir David nugget. It turns out that one of the biggest factors in determining how fast a population grows in any given country is women's rights. In places where women are treated as equals – allowed to make their own choices about education, careers and, importantly, birth control – birth rates are levelling off. In places where patriarchy is still racing ahead – and women are treated as second-class citizens with limited rights – birth rates outpace death rates and so populations continue to soar.

CHIC FACT: You can get birth control pills over the counter in Pakistan, but not in Canada, the USA or France.

It's ALL of our responsibilities to promote equal rights for everyone, regardless of gender (or for that matter colour, sexuality or creed) and the consequences of ignoring this could be catastrophic for everyone. So sexist men (and women) out there, listen up – if there was ever a cast-iron case for feminism then surely it's this: your misogyny is quite literally killing the planet and yourselves. And if anyone tries to argue with that logic, just tell them they're wrong. Sir David Attenborough said so. Try as they might, they can't wriggle out of that one.

If you thought that the Four Horsemen of the Apocalypse, misogyny, a jellyfish apocalypse and facial scrubs were quite enough to contend with then strap in and brace yourselves because it looks like the cosmic powers that be could be plotting an icy attack on life as we know it. In the words of Game of Thrones:

Winter is Coming

> Still . . . in this world only winter is certain.
>
> A Dance with Dragons, *George R. R. Martin*

You probably know that burning all those fossil fuels in our power stations, factories and assorted modes of transportation is slowly but surely heating up our planet and, if not, then you definitely *should* be aware of it. In case the fact of global warming wasn't enough to contend with (global-warming deniers, you can email your objections to openyourfrickingeyes@delusional.com), now we've *also* got a mini-ice age on the horizon.

How's that possible?
You'd be forgiven for thinking that it seems a bit weird that ice sheets could once again stretch down from the North Pole, right across the Arctic Ocean down through the Greenland, Norway and the North Sea to envelope the whole of northern Europe (including good ole' Blighty) in its frozen grip. That surely hasn't happened since the Highlands of Scotland were a playground for the woolly mammoth in 10,000 BC? Why in the world would that happen now in a climate of global warming?

Against a backdrop of nature documentaries and news lamenting the melting glaciers, with images of poor, stranded polar bears floating helplessly out to sea on their shrinking ice rafts, and calls for global powers to slow these rising temperatures pronto, it's hard to see how it's possible for us to be plunged back into an ice age. Yet the possibility of simultaneous global warming and cooling quickly makes sense once we begin to grapple with the fact that what causes these two phenomena are separated by ninety-three million miles. While global warming is caused by our energy-consumption-hungry activities down here on planet Earth, the cause of our imminent mini-ice age is sky-high, originating deep within the huge fiery ball of plasma we call the Sun.

GEEK CORNER: 99.86 per cent of our entire solar system's mass is inside the Sun.

What is the Sun?

The Sun is a huge dense ball of gas. Seventy-five per cent is hydrogen, 24 per cent is helium, 1 per cent is every other chemical element in the universe; the whole of the periodic table is represented. All this gas is so heavy − around 2,000,000,000,000,000,000,000,000,000,000 kilograms − that the Sun holds itself together in that familiar ball-like shape under its own mighty and immense gravitational pull. In fact, the gas ball is held together so tightly that the temperature and pressures are both so enormously high that nuclear fusion takes place, constantly, everywhere. Fusion is the process of squeezing a pair of hydrogen atoms together so closely that, in the memorable words of The Spice Girls, 'two become one'. It takes an awful lot of energy to get fusion to start happening (which is why we've struggled for so long to harness it for power generation here on Earth − but watch this space!); however, once it does get going it can release a phenomenally large amount of energy. So much so that nuclear *fusion* makes its nuclear *fission* cousin − the stuff of nuclear power stations and atom bombs − look positively wimpy by comparison.

GEEK CORNER: Efforts to harness the power of fusion in power stations down here on Earth could quite literally save the planet by producing all the energy we need, but with zero nasty pollutants. The great promise that fusion holds has eluded us for decades, but now, finally, we're getting tantalisingly close!

In the process of squishing two hydrogen nuclei into a single helium nucleus an incredibly powerful, vastly energetic burst of particles and electromagnetic radiation is emitted in every direction. In one particular direction a portion of this strikes planet Earth, its light and heat ultimately allowing all life to exist. So the next time you close your eyes and see the sunlight illuminating the backs of your eyelids, feeling its heat on your skin, give a little nod of reverence to the awesome power of nuclear fusion.

CHIC FACT: If the Sun disappeared we wouldn't know for a full eight minutes.

So far, so impressive, but what's all this got to do with an ice age? We know you can't wait to get to all the mammothy-good stuff . . .

Ever Heard of a Sunspot?

The constant release of electromagnetic energy from nuclear fusion in the Sun results in some exotic and complex interactions between competing magnetic fields, which push and pull against each other in such a way that sunspots are produced. These sunspots block currents of heat from under the Sun's surface, making them cooler than the rest of the Sun. However, all around these sunspots are much hotter patches, spurting out huge amounts of gas and powerful magnetic fields in coronal ejections and solar flares. Admittedly these might sound like the kinds of things our dads wore in the seventies, but they are the real deal. The stunning looking fountains of plasma arc out from the sun's surface in huge plumes, chucking huge amounts of *extra energy* our way.

With all this in mind it's not surprising that the number of sunspots per year affect the Earth's temperature. In years of sunspots aplenty the Earth heats up. In years of fewer than usual sunspots the Earth cools down. Thanks to thousands of years of astronomers keeping an eye on things up there we've long known that there's a rhythm as to how many sunspots occur in any given year and it's impact on global temperatures. What science has only just managed to crack is the ability to model what's going on deep inside the Sun to better predict future sunspot activity.

OK, OK, but where's the woolly mammoth?!
Well, that all depends on:

What's the Sun Gonna Do Next?

By using mathematical models, sunspot activity over the past few hundred years has been accurately recreated. A professor at Northumbria University and her colleagues recently announced that their latest model has improved upon the original and reaching a whopping 97 per cent level of accuracy (see Geek Corner).

GEEK CORNER: In the latest sunspot model two dynamos are accounted for — one rotating deep inside the Sun and another rotating independently closer to the Sun's surface. When they point in the same direction the Sun produces more sunspots than normal. When they point in the opposite direction the Sun produces fewer sunspots than normal, because each cancels out the effects of the other.

Unfortunately for us, their hi-tech crystal ball for sunspots now shows that the number of sunspots is likely to dwindle rapidly between 2030 and 2040. What does that mean for us earthlings? Well, with solar activity plummeting by up to 60 per cent it's gonna get awfully chilly again for the first time for 370 years.* If it's anything like the last mini-ice age in Britain in the seventeenth century (you may be surprised to know we've have many mini-ice ages since the famous Big Freeze 11,000 to 12,000 years ago) it will get so cold we'll be able to ice-skate on a frozen River Thames.

And that's when we finally get our woolly mammoths, right? WE WANT MAMMOTHS!

Well, possibly, because scientists specialising in cloning are controversially thinking about bringing the extinct beasts back using some ancient DNA found in a frozen mammoth's thigh bone and crossing it with an elephant's, but that's a whole other ethically fraught story.

In the meantime, these icy future times may be uncomfortable for the Sun worshippers among us, but at least it will give us a little respite from the never-ending rise in global temperatures. And what's an extra scarf and hand muff between eco-warriors? The trouble is that once the sunspot activity gets back to normal the Earth will heat up once again, melting all that newly formed ice, meaning that the dark imaginings of Kevin Costner's famous damp squib *Waterworld* could actually become a horrifying reality.

What can we do?

---

* During the seventy years between 1645 and 1715 astronomers noted a huge reduction in sunspots, known as the Maunder Minimum, which plunged the Earth into a mini-ice age.

Sadly, there's little we can do to change the behaviour of the Sun, so in the closing years of the 2020s we would strongly recommend stocking up on thermal underwear and woolly jumpers. Maybe even invest in a flock of sheep. Then, towards the end of the 2030s, you'll probably want to buy some inflatable rafts and water purification equipment so that when all the world's rivers burst their banks once the ice caps have remelted – completely this time – you can sail off safely into the wide blue yonder, pet woolly mammoth and all.

We've looked at some of the things that may be heading our way, y'know, the fun stuff like mini-ice ages, plastic seas, jellyfish takeovers and mass extinctions. But what about a threat much closer to home, and one that could have a much more immediate impact on our existence?

Internet Overload

If you're anything like us you probably spend far too much of your day on the internet, looking up all sorts of useful and not-so-useful stuff, checking your Instagram, refreshing your Twitter feed and watching endless videos of kittens on roller skates.*

CHIC FACT: 250 million people currently use Twitter, and over 500 million tweets are sent a day.

Modern man and woman don't even have to bother remembering information any more because, ever since the advent of smartphones brought the internet with us wherever we happen to be, we can access any info we need at the touch of a button. Anyone else remember when you actually had to go to a library to find something out?

History may reflect on the beginning of the twenty-first century as the period in which, in the majority of the developed world, *homo*

---

* Kitten videos have actually been put to the test in surveys and evidence has accumulated to suggest that they really do have a statistically significant ability to make us feel happier – as if we needed a study to prove something so obvious.

sapiens started mainlining information during every waking minute. Suckling at the generous bosom of YouTube, raiding blogs for gossip and pearls of wisdom and trawling online for nuggets of knowledge. For a growing majority of us, these – and many more players in the seemingly limitless supply of sources of new information – have been incorporated into our *daily* digital diet. Internet addiction* is a very real issue.

So can you imagine what would happen if Google started charging 10p per search? There would be uproar! 'Outraged Andy' from Beaconsfield would be just one of a furious clan of marauding parents proclaiming their disbelief, moaning incessantly about their teen-aged kids running up thousands of pounds-worth of Google bills researching their geography coursework (or, more likely, chatting on Facebook). 'How could the government allow this to happen?' 'The Prime Minister must go! *Vive La Revolution!*'

> **GEEK CORNER:** In 2015 internet-related energy con-sumption (think smartphones, computers, server farms) accounted for 8 per cent of total UK power output – equiv-alent to the output of three nuclear power stations. By 2020 the current capacity of optical fibres and switches will have been reached so internet use may well have to be rationed in some way anyway, before the lights start to dim!

Sounds ridiculous, right? But, bitter pill as it may be to swallow, internet shortages may soon become a reality, and our only solution may be to ration it. Or charge for the privilege to reduce the strain on resources.

'Why?' we hear thousands of Snapchat loving teens shout in perfect harmony.

There is something called the 'technosphere' which, despite what you might think, sadly has nothing to do with 180 bpm minimalist electronic music (although it *would* make a good band name!). Broadly speaking, it covers all man-made technologies with a particular focus

---

* In 2014 there were more than 400 internet addiction centres in China alone.

on the resources they consume – electricity, raw materials, space, connections, as well as the human effort expended to build and maintain it all. So, since this lovely word has been invented we can now talk about all these demands under the umbrella term: the technosphere.

CHIC FACT: In 2013 the most common 'what is …?' question searched for on Google was 'What is twerking?' Cheers, Miley! The most watched video on YouTube (at the time of writing) is still Psy's 'Gangnam Style'. Hmmm … maybe internet rationing isn't such a bad idea after all.

Every year computers become more and more powerful.* At the same time internet bandwidth has been getting broader and broader. The combination of these factors means that more digital information criss-crosses our globe than ever before. If we could visualise it, it would be like the greatest laser show the world has ever seen, with seemingly infinite fibrils of light zooming every which way in a spectacle that would make Daft Punk green with envy. However – and here's the problem – in the UK and elsewhere, our internet habits have led to such a vast expansion of the technosphere that we are rapidly approaching the point where demand for electricity to maintain it is outstripping supply. Assuming the power demand of the internet continues to double every four years or so, if all else remains the same the internet will be consuming all the UK's electricity by 2030. And in the current economic climate, we as a nation probably aren't going to be able to increase electricity production quickly enough to keep pace. Which means only one thing: internet rationing (*shudder*).

Tim Berners-Lee famously declared that the internet would be *free for all* when he originally invented it and during his London Olympics 2012 appearance he went on to tweet: 'This Is For Everyone.'

---

* In 1965, the man who would go on to co-found Intel – Gordon E. Moore – famously stated that the number of transistors in a circuit would double every year. Ten years on and he revised his prediction to doubling every two years, which seemed to do very nicely: more than 50 years later it's still in pretty good shape.

However, the reality is that we may soon find ourselves being issued with a month's worth of digital ration cards.* When we end up reaching our limit, we'll look back on whole nights spent binge-watching Netflix, exhaustively searching the internet for cute dog memes or cyber-stalking our old flame's Insta-pics, as halcyon days of yore. Many of us glaze over in ecstasy at the thought of going back to the free-love days of the swinging sixties, but the people of the future may well look back jealously at our current era as an unrestricted orgy of digital ecstasy.

If you've only just managed to get your head around the fact that soon our internet usage may become heavily rationed then get ready for another little shock, because something even closer to our caffeine-loving hearts could be under serious threat.

Coffee Crisis

At rush hour in any of the world's major cities you'll see swarms of people scurrying to school or work protectively clutching their morning cup of coffee. It's almost as if they're proud to admit their addiction to the world's number one, bona fide, smart drug, one that your authors have a serious soft spot for. And it's neuroprotective,† as it happens. It's impossible to escape it – every modern high street these days seems to have been invaded by a surplus of coffee shops,

---

* A meeting of leading academics at London's Royal Society in May 2015 discussed the very real possibility of internet rationing in the near future.
† Three to five cups of coffee per day is considered a moderate dose which has been associated with a neuroprotective effect: people who enjoy a moderate daily consumption seem to develop Alzheimer's and Parkinson's diseases later, on average, than their non-coffee drinking counterparts!

each competing for the right to feed our need for that brain-jolting injection of delicious, bitter-sweet black gold.

It's not hard to imagine how this tots up to a two-billion-cup-a-day habit globally. The winning bean? *Coffea arabica* – by far the world's number one variety. The smooth and complex taste of the *arabica* cherry – that's what pros call the *arabica* bean, in their best Loyd Grossman voice, no doubt – is vastly preferred to its much coarser and more bitter cousin *Coffea robusta*, which is used almost exclusively for instant coffee. Hundreds of years ago, cuttings snipped from a few *arabica* plants in the north-western Ethiopian highlands were distributed all over the world to supply the steadily increasing demand for coffee. Any place that maintained the year-round 18–22°C that *arabica* thrives in was ripe to play a role in the colonial expansion of the mighty A-bean.

So all those exotic Colombian, Jamaican or Indonesian blends we enjoy today are actually the descendants of African immigrants (a bit like us!), criss-crossing continents and oceans to settle in Asian and Central and South American soils. The trillions of cups of coffee we collectively consume as the years go by all derive from a handful of ancient plant cuttings taken from Ethiopia. In the words of Michael Caine, 'not a lot of people know that'.

However, we've got some bad news for you, so make sure you're sitting down and have a big cup of the good stuff in hand to revive you, because – deep breath now – your daily coffee may soon become an indulgence you can no longer afford. What! Why? How? Really!? Yup, 'fraid so, oh caffeine-loving friends. Unfortunately, we humans are on the verge of a global coffee crisis.

At the very core of this crisis is the vanishingly small gene pool in our coffee plantations all over the world. A single coffee plant cutting was more often than not all that was used to start the overseas colonies of *arabica* plantations, so all generations that followed shared the same vanishingly small gene pool, which limits their capacity to cope with all sorts of hurdles, like disease-causing agents and extreme weather.*

---

* The same goes for us humans. Have babies with anyone but your siblings or other close family and you can be pretty sure that your offspring will enjoy all the benefits from a much wider gene pool: such as feet that aren't webbed and an IQ greater than sixty.

With changing climate placing unprecedented new pressures on *ara-bica* plantations, crop failures are happening all over the world. These inbred varieties simply don't have the genetic reserves found in the population as a whole that would have helped their great-great-planty ancestors survive challenging climate changes in the past. Add to all this the fact that global warming is playing havoc with the *arabica's* reproductive system and, to put it plainly, our favourite little bean is getting a right old drubbing.

GEEK CORNER: The Achilles heel of *Coffea arabica* is its teeny tiny gene pool. The lack of genetic diversity leaves the organism much more prone to disease. In the 124 varieties of coffee that grow in the wild, each has a bunch of genetic material not shared by other plants. Some of this unique DNA will contain the code for special proteins that helped the plant adapt to unfavourable changes in climate, the attack of a pest or an ill-timed downpour at some point in the distant past. This enabled their ancestors to survive while other ancient wild coffee varieties perished. A certain proportion of any wild population can usually survive any given challenge, but no *individual* plant will contain all the genes that protect against all possible threats, which is instead held by the entire population. Hence the problem with starting out with just a few cuttings.

> **CHIC FACT:** Coffee has been banned five times in history, most recently in 1777 when Frederick the Great of Prussia outlawed it because he felt it was interfering with his people's beer consumption.

Don't panic. Take another deep breath. There *is* an alternative. You can always develop a taste for the incredibly bitter *robusta* variety instead. As its name might suggest, it's a hardy little bugger, much more resistant to pests and climate variation than its wimpy little cousin *arabica*, but the problem is that for many of us, it's just not quite as tasty.

Or, if you don't fancy a bitter cuppa in the morning, the science world has a clever plan. Now that we've realised quite how tiny the *arabica* gene pool is, and therefore how vulnerable these plants are to climate change, scientists across the world are hungrily pursuing the possibility of crossbreeding this variety with some of the other, tougher varieties of coffee plant. There are many other varieties besides *arabica* and *robusta*. The hope is that they'll be able to create a super-hybrid which produces the same delicious *arabica* taste we love, yet also manages to thrive in climates beyond the preferred 18–22°C, even when perpetually pestered by various pests.

In the meantime, they've got another quick fix up their green-fingered sleeves: grafting the tasty *arabica* bush onto the tough *robusta* roots – like some kind of caffeinated Frankenstein's monster. That should keep us going for a few more decades before the pure *arabica* flat white becomes the exclusive preserve of billionaire oligarchs and we have to blag an invite to Roman Abramovich's to even get a whiff of the sweet, sweet stuff.

> **CHIC FACT:** Coffee is actually a fruit, originally eaten as a food in the form of a snack ball made from coffee berries mixed with fat. Mmmm, delicious.

It's pretty clear – our planet is in a spot of bother. But rather than sink into a Sartre-esque pit of nihilistic depression, many are doing their very best to think up ways to protect our planet and the innocent species that share it with us. And, as the saying goes, prevention is better than cure. Here's one of our favourite examples.

Pimp My Rhino

American rapper Xzibit is sitting in a garage, waiting for his next customer for MTV's new series: *Pimp My Ride: South Africa Special*.

The tunes are blaring, the heat is shimmering above the tarmac and in walks Safari Mack, head gamekeeper at Kruger National Park. In his hand is a long skinny rope and on the other end of it plods in a huge, beautiful, grey beast:

'Howzit, bru! I'm here for the TV show and I brought this fella along with me. So, come on then . . . who's gonna Pimp My Rhino?!'

You can see the scene unfolding, with the souped-up rhino eventually emerging in the big reveal wearing roller skates, a Tudor ruff and a new diamond-encrusted horn tip sparkling in the sunlight. But as ridiculous as this fantasy scenario may be, rhino horns in real life have been getting a bit of a makeover, and it's not for TV – but for life-saving reasons.

CHIC FACT: The word rhinoceros literally means 'nose horn'.

The human desire for carved rhino horns – alongside various myths about their medicinal benefits that have been flying around for the past two thousand years or so – is now driving the remaining five species of wild rhinos into extinction. Rumours of health benefits from consuming shaved or ground-up rhino horns persist even today, despite zero evidence to suggest they have any medicinal value whatsoever. It's said they cure everything from fever and hangovers, to rheumatism and cancer – one twenty-first-century newspaper article ludicrously claimed that a Vietnamese politician had been cured of cancer after drinking crushed rhino horn, resulting in a huge rise in poaching. The misplaced beliefs of millions in Asia is leading directly to the massacre of Africa's most impressively horned majestic beast. Even by the stupendously odd standards of weird science that is a pretty strange set of circumstances. Rhino poaching has shot up recently, rising drastically from thirteen in 2007, to a few hundred per year in 2010, to over a thousand per year since 2013 – and that's in South Africa alone. Already in the past decade we've seen the West African black rhino become extinct, while Javan and Sumatran rhinos are now critically endangered. It's nothing short of a travesty.

CHIC FACT: Traditional Chinese medicine has prescribed rhino horn as an ingredient in cures for fevers and liver problems for almost two thousand years. As both rhino horn and our own fingernails are made from a protein called keratin the medicine is likely to have no more medicinal benefit than chewing your own fingernails. Alas, the complete absence of evidence does little to stem the murderous flow.

What's more, after poachers have chopped off their horns, the poor rhinos' bodies are routinely left for the vultures and hyenas to gorge themselves on beneath the scorching African sun. To put it plainly, thanks to some of mankind's bonkers beliefs, not to mention a complete lack of respect for wild creatures, our beautiful horned friends are now in serious trouble.

However, many of the world's most brilliant minds are thinking up ways to combat the threat from poaching. Several nifty and innovative solutions have already been tried out to pimp out rhinos' horns with some technological protection measures. These include horn burglar alarms, horn-cam CCTV, and our favourite remedy – filling the interior of the horn with a nasty poison which is harmless to the rhino and painting the outside pink. Poachers see the pink and wisely steer clear of the toxic horns knowing that they will have been injected with parasiticides. Not only would handling it cause their hands to break out in a rash, but the Asian market would soon stop buying from them when their customers starting getting sick after consuming it.

We *love* this last approach. Firstly, because it's incredibly bold and the shocking pink horns make the rhinos look punkilicious. But mostly we're fans because it's a solution that takes a 'helicopter's eye'

view on the overall situation by relying heavily on the well-known 'bush telegraph' of local gossip.

CHIC FACT: In Vietnam rhino horn fetches prices in excess of $10,000 per kilogram – that makes rhino horn worth more than its weight in gold.

Once the gossip of the pink-toxic-horn-tampering has spread from village to village, poachers know to steer clear of the whole area. There's no point wasting precious hunting time on an area loaded with poison-horned rhinos. To help ensure that the message spreads, the Rhino Rescue Project behind this ingenious move even goes as far as inviting the local villagers along to watch the toxin-injecting and horn-pinking process, encouraging them to take and share photos on their mobile phones to ensure that everyone knows: pink means poison; poison means no profit.

While drilling a hole in the unconscious rhino's horn and filling it with the world's most powerful itching powder (and then drawing attention to what you've done by making the horn look like it's dolled up to go techno raving) might not be everyone's idea of a fun afternoon, it is an ingenious solution to a previously unsolvable problem. As for *Pimp My Rhino*? That one's coming to a TV nowhere near you, no time soon.

Finding new and ingenious ways of protecting other species is a huge part of our struggle to look after our planet, but when it comes to looking after ourselves we need to start conserving the world's resources. When trying to cope with the fallout of this seemingly inevitable apocalypse of ours it makes sense to look towards technologies developed for use in the developing world. As developed as our own countries might be at the moment, just one stray comet, super-volcano or magnificent epidemic could set us back a few centuries. At which point poo-lectricity could end up being a life-saver for us all.

Poo Power

Getting our electricity from 'poo power' may not sound all that enticing a prospect, but scientists have developed a way to do it, and guess what? It kind of makes sense. After all, animal poo in the form of manure is typically used to help crops grow, so why shouldn't our waste be put to good use too?

The Bristol Robotics Laboratory is leading the charge in this field and has created a robot – called EcoBot III – that collects its own food and water from the environment, artificially 'digests' it and then essentially craps its robo-mess into a litter tray. The nutrient-rich waste is then transformed into energy using a relatively simple set-up. Microbial fuel cells work a bit like batteries, using bacteria to metabolise the robo-faeces and turn it into electricity (see Geek Corner), which is used to power the robot itself. It even has a little fly-trapping hat to keep pests at bay. Nifty!

GEEK CORNER: Just like batteries, microbial fuel cells (MFCs) contain an anode and a cathode electrode, separated by a thin membrane but each connected to an external circuit. Bacteria at the anode end digest organic matter fed to them (such as sugar, or poo). Normally, in the presence of oxygen, the microbes would then respire and produce carbon dioxide and water. However, as there is no oxygen inside the MFC (known as anaerobic conditions), the bacteria metabolise to produce carbon dioxide, protons and electrons instead. The protons pass through the membrane from the anode to the cathode, while the electrons pass through the anode to the cathode via an external circuit. The difference in voltage between the anode and the cathode is the process that creates a current.

EcoBot III was created in a neat collaboration with Wessex Water, a sewage company whose expertise was invaluable in developing the world's first waste-to-energy bot. Their joint aim is to use human waste from toilets for energy. We can imagine how the negotiation went: 'you provide the logs, we'll provide the 'bots, deal?'

But, really, what's the point? Surely we can get energy from other less vom-inducing sources? We could, but treating the huge amounts of waste processed by our sewage systems every single day uses up an awful lot of energy in itself. The ultimate goal today is to scale up EcoBot III technology to the point where sewage plants are powered using energy extracted from the very waste they're treating. Self-sufficient, renewable, efficient and cheap energy for sewage waste treatment? Sounds like a winner to us.

Deriving energy from human waste is nothing new – the Assyrians used biogas from their poo to heat up their bathwater way back in the tenth century BC – and today Nairobi is already taking advantage of 'poo-power'. In the slums of Nairobi and other areas of Africa, 'flying toilets' – otherwise known as bags of poo – are commonly flung out of the windows of various ramshackle houses. Kenyan scientists and aid workers were determined to put this readily available power source to good use, and at the same time reduce the hazardous waste on the streets, so they installed purpose-built 'bio-centres' to house the collected poo, gather the methane gas it emits and sell this 'biogas' for cooking or to power hot showers. However, unlike the biogas digesters commonly used in Africa to convert human waste into energy, the microbial fuel cells are very efficient (85 vs 35 per cent).

CHIC FACT: In 2014 the first British 'poo-bus' went into action, powered by the biomethane collected from human faeces at sewage plants. And the poo-bus isn't the only human-waste fuelled transport. In 2010, the 'bio-bug' was developed, converting a VW Beetle to run solely on biogas, sourced from local sewage plants. The Beetle could run for 10,000 miles at speeds of up to 114 mph using the human excrement from just seventy homes. Brings new meaning to 'Dung-Beetles'.

Meanwhile, both Sweden and Germany are using biogas from human waste to power some of their national grid. If the UK clicked on, we could fully power 350,000 homes a year just using the energy derived from poo. So should some awful cataclysmic event one day rob us of the plentiful source of electricity we've come to rely on, EcoBot III technology may eventually come in very useful. As long as it comes with a lifetime's supply of *eau de toilette* . . .

Besides, 'doo-doo-energy' is never going to run out as long as we're knocking about on the planet (unlike other energy resources, which we're fast depleting). With 100–250 grams of faeces produced per person per day and with approximately seven billion people on Earth, that's somewhere between 250 and 640 billion kilograms of human waste produced every single day (and that's not even beginning to include that produced by other animals!). No wonder many are now starting to see poo as the ultimate source of never-ending renewable energy.

Number twos aren't the only source of human waste being used to generate energy; number ones are getting a look in, too . . .

Wee-lectricity

The simple act of going for a wee could soon generate precious electricity thanks to the Bristol Robotics Laboratory pioneering the UK's research into urine power, the same team behind the poo-powered robot EcoBot III.

> **CHIC FACT:** With over seven billion people on Earth there's approximately 10.5 billion litres of human urine produced every single day. Given that it can easily be harnessed to generate much-needed electricity, it seems an awful shame that this daily supply of 4,200 Olympic-sized swimming pools of wee is going to waste.

In 2013 there was a breakthrough when the team managed to power a mobile phone using energy derived from human wee alone.

The phone could call, text and even browse the web, which would make for an arresting Tweet: 'brought to you by my own piddle!'

As with the poo-powered robot, the wee-powered phone runs using microbial fuel cells which work a bit like batteries (as mentioned earlier). Special bacteria metabolise the organic compounds in the urine and release electrons, passing them through an electrical circuit to generate electricity.

> **CHIC FACT:** The wee-powered mobile phone needed just twenty-four microbial fuel cells to function, whereas the pee-powered toilet lights required 288.

Urine-tricity is now being trialled in a groundbreaking study as the sole source of energy for lights in a new prototype toilet. It's being tested at a British university (in collaboration with Oxfam) with the noble aim of making going to the loo at night safer for vulnerable women in places where electricity is in short supply, such as disaster zones and refugee camps. The idea is that the raw material for the pee-power is provided by the very people who need light when they come to use this futuristic loo. Again, microbial fuel cells generate the electricity that is stored in a capacitor to power the lights. The whole thing is wonderfully self-sufficient and totally renewable. At £600 for the full set-up it could well be the best mixture of affordability, green technology and humanity-boosting science we've seen.

And if mobile phones and toilet lights can be powered by human wee, then so can anything else that requires electricity. More than a billion people, a seventh of the global population, across the world have no access to electricity today, and yet almost everyone across the world has access to urine – so pee-powered electrics make a whole steaming heap of sense.

CHIC FACT: A Sardinian researcher has managed to harness 'pee-power' to run a car, believing it could one day replace petrol – which would be a great help in reducing both rising petrol bills and oil-related warfare. Wee-mobile, anyone?

Human beings are clearly making some bold moves to prevent the destruction of our planet and to think up some clever ways to generate power – but what about the most precious resource of all? The key to our survival – drinking water.

Post-Apocalyptic Drinking Water

Let's imagine you are one of the lucky few who, either through excellent planning or sheer blind luck, managed to survive an apocalypse that wipes out 99.9 per cent of the Earth's population. What would your first priority be? Hotwire a sports car and take it for a spin? Break into a mansion? Help yourself to a soothing Jacuzzi and then head to their walk-in wardrobe to try on some flash togs?

Such distractions would no doubt be great fun for a while, but it soon would dawn on you that life as you know it had been fundamentally altered. With so many wiped out there may be no functioning infrastructure to bring electricity to your power sockets, no agriculture to bring fresh food to the shelves of your local supermarket and, without functional water treatment plants, you'd have to learn pretty fast to avoid tap water if you didn't want to accidentally poison yourself.*

Given how important water is to survival you'd be well advised to ensure you had access to a clean source. Assuming you don't have an unspoiled well to hand, you'd need to find a way to purify dirty

---

* For more on this we highly recommend a book called *The Knowledge: How to Rebuild Our World from Scratch* by our favourite astrobiologist, Lewis Dartnell, who we interviewed in our apocalypse special Geek Chic Weird Science podcast.

water, or better still extract it from thin air. Luckily for us imaginary apocalypse survivors, present-day innovators are already thinking about how we might get clean drinking water to people living in such unfortunate circumstances right NOW.

Let's be honest, while most of us in the developed world are lucky enough to only to have to imagine this hypothetical apocalyptic time when fresh water is no longer readily available, others in the developing world are not so fortunate. As we've seen in many drought-ridden developing countries over recent years, a safe supply of sanitary water can mean the difference between life and death, making sourcing fresh drinking water an absolute priority. Many scientists are now focused on creating solutions that would work in the developing world and potentially save millions of lives today – and should the apocalypse ever happen these could come in very useful for the survivors, too.

Here are just two of the clever ways scientists and engineers are currently striving to provide clean, drinkable water.

Water from Thin Air

You know that feeling when you're halfway through a really long and tough bike ride – you're feeling exhausted, sweaty and very, very thirsty. You reach down to your water bottle in anticipation of the gorgeous thirst-quenching drink you're about to enjoy, only to put the bottle to your lips, tilt it up and then . . . nothing. Dry as a mutha' humping bone.

It happens to the best of us, no matter how well prepared we think we are, and we don't know about you, but we spent many a childhood daytrip fighting our siblings over those last few precious drops. As we played tug-of-war over the water bottle, little did we realise how close we were to an invisible supply of fresh water pressing in on us from every direction: water vapour suspended in the very air that surrounds us.

The only problem is how to convert this water vapour from a gas suspended in the air into liquid water droplets and then collect enough of them to be able to gulp them down as a refreshing drink. Luckily for us Kristof Retezár, designer of the Dyson Award-winning Fontus, reckons he might have found the solution with his nifty little auto-refilling water bottle for cyclists.

The secret of this contraption is condensation. When warm, humid air makes contact with a cool surface, it is converted into liquid form. Think of all the liquid water droplets that appear on your mirror when you hop out of a hot, steamy shower – water vapour is cooled by the glass to become liquid water and *hey presto!*, you've got something refreshing to lick off of your reflection – literally out of thin air.

> **CHIC FACT:** The practice of creating water from thin air isn't a new one. Some cultures in Africa and South America have been using condensation methods to get drinking water from the air for more than two thousand years.

The bottle uses similar physics in a little device at its centre, known as a Peltier Element. It is powered by solar energy to give it a hot bottom and a cool top, causing water vapour to rise from the hot bottom upwards (because hot air rises), then hit the cold top and form condensation droplets. These then flow down a pipe and collect at the bottom of the container. It's a relatively slow process, with drops collecting at a rate of one per minute, but the water bottle does fill up by about half a litre every thirty minutes, which the cyclist can then sup down to their heart's content.

Once we've mastered the art of creating drinking water from thin air on a small scale like this, there's no reason we can't put our minds to scaling it up to something much larger. With over a billion people around the world living without access to safe drinking water, this simple invention could one day evolve into something that saves innumerable lives. And would no doubt come in mighty handy

should life as we know it be disrupted by war, or some apocalyptic event ... something like a devestatingly large comet touching down on Earth.

GEEK CORNER: The Earth's atmosphere contains around 13,000 km^3 of fresh water – equivalent to 5.2 trillion Olympic swimming pools*. That's an awful lot of bottles of Evian floating around unexploited.

Clean Water from Crisp Packets

If you thought that empty crisp packets were only good for putting in the oven, watching them shrink and then turning them into tiny badges, think again. Crisp packets can actually SAVE LIVES today. And this solution could also easily save your skin on a post-apocalyptic planet Earth.

Estimates suggest that, at present, 1.5 million people die each year from drinking contaminated water. Ninety per cent of these are children, which is mind-boggling when you think that so many of us take for granted turning on a tap for unlimited clean water to come gushing out. Often, the only water people in the developing world have access to is contaminated and unfit to drink without some method of purification. For example, someone might typically be able collect rainwater in a vat, but this invariably ends up infected by microbes or other pathogens. Such impure water sources ultimately cause sickness and even death if not treated properly before consumption.

CHIC FACT: The World Health Organization has calculated that 1.8 billion people around the world only have access to water that is faecally contaminated.

---

* Any Olympic swimming pool contains 2.5 million litres of water (i.e. 2,500,000 litres), one cubic kilometre is 1 trillion litres of water (i.e. 1,000,000,000,000 litres), 13,000 km^3 is thus equivalent to 5.2 trillion Olympic swimming pools (5,200,000,000 litres).

The usual methods of purifying the water can be complicated or expensive, but a team of scientists and engineers in Australia have come up with an ingenious crisp-based solution.

Part of the beauty of their water treatment system is that it can be built by anyone with very limited materials – you simply take a large, clean, semi-cylindrical vessel (a thoroughly-cleaned oil drum cut in half should do the trick, but they used a glass one), then line the surface with some inside-out empty crisp packets ensuring that the silver side is facing up towards the sky. Once you have done this, you hold it up towards the Sun in a frame made out of plywood and fill it with water. The crisp packets coating the half-cylinder redirect the Sun's rays and focus its light energy into the water. Vitally, the UVA part of the sunlight is concentrated as it is reflected onto the water tube, creating enough energy to form 'reactive oxygen species' in the water – chemical substances that zap the bejesus out of the bacteria, permanently damaging their DNA, preventing them from multiplying and ultimately killing them all off.

The hot-shot team from the University of Adelaide demonstrated that 40 litres of unclean water, riddled with *E. coli* and other types of nasty infection-causing bacteria, was rendered safe enough to drink in only thirty minutes. Just half an hour in the device was sufficient to reduce levels of pathogens in the water to undetectable levels.

It's incredibly simple, costs just £40 in materials to set up and works so well. Not only that, but once someone knows how to do it, they can keep sharing this blueprint with everyone else around them. The best part? The altruistic Aussie researchers decided to waive their right to patent the contraption, meaning anyone, anywhere, can use it to save lives.

So, next time someone tells you that crisps aren't good for you, you can tell them to back down, because you're saving the world, one Monster Munch at a time.

Final Thoughts

So, peeps, what have we learned? Other than that we are in a whole heap of doo doo and it's all our fault . . .

Well, human beings may be a bit of a scourge as far as the Earth's concerned, but we are ALSO the key to its ultimate survival.

Focusing on clean, renewable energy sources, cutting down on pollution, generally respecting our planet and the other creatures that inhabit it and, perhaps most importantly of all (if Sir David Attenborough is to be believed), promoting equal rights for women are things that each and every one of us can strive to do. Vital to this will be us ordinary folks putting pressure on our governments to make big legislative changes, which can have huge, rippling, positive effects on the rest of the world. As individuals there is little that each one of us can do to help, but many voices are greater than one, and collectively we can raise awareness and motivation to the point where things do actually change.

In the meantime, enjoy life while we have it and try not to stress *too* much about the fate of our species. We are only passing through on this world for a vanishingly short time in the grand scheme of things, and if that concept doesn't cheer you up much, then just remember our clean water from crisp packets story may have given you the perfect excuse to eat a shed load of Hula Hoops without feeling guilty. Suddenly life doesn't seem so bad after all.

8

Food, Glorious Food

One quarter of what you eat keeps you alive.
The other three-quarters keeps your doctor alive.
Hieroglyph found in an ancient Egyptian tomb

We humans have a strange and often ambivalent relationship with food. We collectively spend billions of pounds on our gastronomic fixation, worshipping celebrity chefs like the rock gods of yesteryear. Yet we also seem to be in a continual battle against food, constantly switching between *this* diet and *that* diet as if they're the answer to our culinary confusion.

Newfangled heroic superfoods are pitted against new food super-villains almost every single week – only to be trumped by yet another bunch of candidates soon after. We're constantly bombarded by the REAL TRUTH about food. Fat is the devil-child one month, but then, wait, sugar is the enemy the next. Soya milk is the saviour of mankind (and cowkind) one minute and the next it could be cruelly ruining all your chances of having babies. Then, of course, there's sliced bread, (though as we all know nothing has ever really been the 'best thing since') which is now getting a kicking for a lack of nutrients – instead it's packed with sugar and salt to give it flavour. No wonder everyone's feeling so damned confused.

There's not much we can do to halt this assault of conflicting information. As with everything in the science world, new theories will continue to emerge, be disproven and then fall back into favour a few years later. Real progress towards any definitive answers in science tends to move to the rhythm of decades, not years. Yet any eye-catching little piece of the puzzle – evidence supporting or refuting a nutritional idea – is amplified by the media and put forward as a new, indisputable fact. 'We've cracked it, kids, blueberries really *will* make you live forever, be smarter, grow taller and look just like Scarlett Johansson. Promise.'

It would make much more sense to discuss new evidence in relation to all the previous research on the food in question; something along the lines of: 'this study seems to support the idea that blueberries have some positive health-giving qualities, and how this fits into the overall picture is . . .' Unfortunately for us consumers, this more balanced yet slightly snooze-worthy approach doesn't sell newspapers and magazines, so we're stuck with the attention-grabbing, but often misleading, headlines.

One of the problems with all these controversial foodie 'facts' is that they can start to make even the healthiest person paranoid about everything they're shoving in their gob. People try so hard to treat their bodies like a temple, but end up becoming confused by all the mixed messages that they give up altogether and descend into a culinary free-for-all, stuffing their faces on yet another binge.

We'd do much better to adopt a more balanced approach and take any new advice with a tiny pinch of salt, remembering all the while that moderation is usually key to not falling foul of food's darker side, and that a small dose is very unlikely to damage your health too badly. After all, food is one of life's greatest pleasures and we need to eat regularly if we want to survive, so we might as well make the most of it.

With that in mind, we've dug out some of the funniest, tastiest and juiciest stories from the world of food science to tickle and horrify your taste buds in equal measure. Over the following chapter we'll show you how some of the food and drink you thought was 'naughty' – like booze, coffee and junk food – can actually be 'nice' for your body and brain. We'll also take a peek at some of the more recent bizarre tales from the ever-changing world of food, debunk some of the sneakier headline-grabbing revelations and share some of the latest, most surprising new studies. We'll even throw in a tale about the narcotic escapades of one of the world's most delicious fish, so get ready to see food in a whole new light.

Cheesy Heroin

We are serious cheese lovers, bordering on the obsessive. We confess to sometimes eating blocks of cheddar as if they were apples. There are times when we've even Googled 'cheeseoholics anonymous' in the hopes of finding support groups to help wean ourselves off it – while absent-mindedly shoving cheese strings into our mouths, of course. Perhaps you even delight in some cheesy goodness yourself, with a fridge permanently brimming with Brie, Jarlsberg, Emmental and Caerphilly. Then there are the pizza worshippers among you. The cheese-on-toast compulsives. The nacho nutters. And the fondu fanciers.

> **CHIC FACT:** When it comes to compulsive eating there's little that tops a cheesy pizza. In a recent study it was crowned 'most problematic' in terms of how much people struggle to control their consumption as measured by the Yale Food Addiction Scale.

Could it be that all this dairy lust goes way beyond the everyday enjoyment of food? Could it be that we are actually *addicted* to cheese?

Recent headlines suggested that we might be, loudly proclaiming that all that soft, warm, gooey cheese is addictive in a way not dissimilar to hard drugs. Now, they weren't necessarily suggesting

that people might start shooting up Dairylea in shop doorways in a vain effort to chase the dragon. More that, by looking at the chemical substances involved, there does seem to be a bit of an overlap between cheese and heroin that could potentially explain why it is so very moreish.

The chief suspect is casein, which makes up 80 per cent of the protein in cows' milk. Once ingested, it is chemically trimmed by our digestive systems into casomorphins, which derive their name from Latin and roughly translate as 'cheesy morphine'. Casomorphins can tickle the opiate system through which drugs like morphine and heroin act, so some thought it was perfectly plausible that tucking in to cheese could trigger a similar response.

CHIC FACT: The word 'morphine' stems from German and French words for the Roman god of dreams, Morpheus, who could appear in people's dreams, taking on any human form. In his true form he was a winged daemon. This seems apt – we quite like to think of the casomorphins as tiny winged wedges of Edam cackling maniacally like the flying monkeys from *The Wizard of Oz*.

One particular study found that when casomorphins were injected directly into the brains of newborn mice there was indeed some evidence of pain relief, albeit at a level just a twentieth as powerful as morphine. Aha! – perhaps Cheeseoholics Anonymous isn't such a stupid idea after all.

The only problem with this reasonable-sounding theory is that in the mouse study casomorphins got into their tiny brains via a syringe. There's little evidence to suggest that casomorphins (which enter our bloodstream via the mouth and stomach) can actually get into our brains from our blood, which of course is exactly where they would need to go if they were to meddle directly with the brain mechanisms of addiction. The blood brain barrier (BBB) is a special layer of defence between brain tissue and the zillions of blood vessels passing through it. A bit like some hulking bouncers at the door of a club, the BBB closely controls who comes in and who goes out. And casomorphins are strictly not on the guest list. No pleading or dressing up can get them past the heavies. Could they sneak in through the back door or under the coat of a fellow opiate receptor? It's possible, but we are certainly nowhere near to proving this yet.

GEEK CORNER: In the fifties and sixties, *Monoamine oxidase inhibitors* (MAOIs) were all the rage among psychiatrists for treating depression and several other mental health issues. By reducing the breakdown of various brain chemicals important for emotional regulation, these drugs were effective in improving mood, but patients on such medication had to beware of the dreaded 'cheese effect'. A good ol' cheese binge would lead to pounding headaches and occasionally much more serious complications. The reason for this is that cheese, particularly aged cheeses like Stilton, contains loads of tyramine. Under normal circumstances the MAO enzyme would break this down before it could cause any trouble, but in those treated with MAOI medications this enzyme is rendered ineffective. All the extra tyramine from the cheese could then swim around their bodies and brains unopposed, triggering the release of dopamine, adrenaline and noradrenaline, causing their blood pressure to rocket.

For the time being the idea that casomorphins in our cheesy dishes are strumming the brain pathways of addiction is still more theory than fact. But as it's the food we find most irresistible, we won't be surprised if it ultimately does turn out to be the case. More research is definitely needed and we would like to be among the first to take this opportunity to put ourselves forward to take part in any future cheese-munching investigations. All in the name of science, of course.

It's one thing exploring whether certain foods such as cheese might trigger pleasure responses in our brains in a similar way to heroin, but what if we told you some foods go one step further? What if actual cocaine ended up on your dinner plate and you were none the wiser?

Coked-up Fish

It's tough being a salmon. One moment your children are happily swimming around, innocently playing in the river weeds, and the next they're all grown up, hanging out by the rocks with some dodgy looking eels and getting high on drugs. Sadly, for the parents of young wild salmon in Seattle, USA, their worst nightmares seem to have come true – their juvenile salmon have tested positive for cocaine.

CHIC FACT: The real term for baby fish is fry – so next time you call someone 'small fry', remember you're really saying that they're a teeny tiny fish.

The sample – a bunch of migrating Chinook salmon and staghorn sculpin caught off the coast of Washington State in an inlet of the Pacific Ocean known as Puget Sound – were actually found to be contaminated by over eighty different drugs including cocaine, anti-depressants, nicotine and caffeine. Did this particular school of fish simply bunk off the *Just Say No!* classes at school? Were their parents too liberal? Or are salmon kids partial to a spot of risky living? The answer – unsurprisingly – is none of the above. The all-too-predictable culprit is mankind.

What transpired from the study is that some of the local population of drug-dependent humans, from nearby cities such as Seattle, Tacoma and Olympia, were popping pills and snorting lines, which meant they then flushed their drugged-up bodily wastes down the toilet. The contaminated water would flow to the local sewage plant, out into rivers and eventually end up in the sea. The poor fishy neighbours of the drug-loving folk were being spiked by the noxious substances still contained in the supposedly 'purified' sewage water and were absorbing them into their tissues. The drugged-up fish were then exhibiting strange behaviours such as spontaneously changing sex and having aggressive outbursts.

No one's completely sure whether it's down to a failure in the local sewage plant's cleaning process or simply that an awful lot people in those nearby cities like to get high, meaning they're overwhelming the cleaning process; either way, it's a pretty big problem. Not only for the wild salmon and other marine life affected but also for the men, women and, most worryingly, children who innocently sit down to tuck into a tasty plate of fish 'n' cocaine.

CHIC FACT: Fish aren't the only creatures getting drugged up thanks to human consumption of psychoactive chemicals. Starling populations in certain parts of the UK are seemingly plummeting and some suspect it's because the worms they like to feast on are themselves chowing down on human waste products from local sewage plants containing low concentrations of Prozac. Whilst you'd be forgiven for thinking the birds would end up enjoying the twin benefits of a more stable mood and cheerier disposition, they are also suffering from some of the lesser known side effects of the world's most popular selective serotonin reuptake inhibitor (SSRI): suppression of both appetite and libido. So they're not hungry, they're not horny and their numbers are dwindling accordingly.

Fish in Puget Sound may be being poisoned by man-made drugs, which eventually end up on our dinner plates and in our bodies but, globally, a major problem is occurring on a far larger scale. Livestock animals around the world are routinely fed drugs, but in their case it's antibiotics, and the long-term consequences could be catastrophic for all of us.

Meat-Munching Mayhem

> 'There may be no other single human activity that has a bigger impact on the planet than the raising of livestock'
>
> *Time*

T-bone steaks, BBQ ribs, lamb chops, chicken fillet, liver-and-bacon casserole, turkey escalope, mutton stew, curried goat, smoked ham, springbok biltong, salami, spatchcock guinea pig and frog's legs. For one of your authors, this is a mouth-watering menu; just a few of the meats he's taken great pleasure in trying over the years. For the other, not so much. In fact, as a non-meat eater, that list made her feel a touch queasy. Other vegetarians reading this may feel the same and by the end of the section their emotions may have morphed into full-blown fury, which to be honest is totally understandable. It must be frustrating for veggies to learn that every human being on Earth is being placed at greater risk of plunging into a perilous superbug-ridden dystopia thanks to the voracious appetites of human carnivores.

Let's explain. Once upon a time, when meat was a luxury, most people would tuck into it once or twice a week at most. Nowadays, the majority of meat eaters have come to expect it once or twice a day. It's got to the point where, worldwide, 285 million tons of meat are produced and consumed every single year. This number is set to rocket in years to come as the global middle class – those who can afford to fund such a regular meat habit – expands at an astonishing rate.

CHIC FACT: Each year 586 million tons of milk, 124 million tons of poultry, 91 million tons of pork, 59 million tons of cattle and buffalo meat, and 11 million tons of meat from sheep and goats is produced. That's an annual total of 285 million tons of meat produced globally.

We already know that there are big problems associated with all this meat munching. Aside from the raft of ethical considerations, the ecological issues are enormous. For example, 30 per cent of the total surface of the Earth is now used either to raise livestock or the grains and other crops that feed them, destroying huge areas of rainforest and other natural lands to make way for it. Livestock farming uses a third of the world's fresh water (cows have to slurp a whopping 990 litres of water to produce just 1 litre of milk) and produces millions of tonnes of ammonia, which exacerbates the acid rain problem. Thanks to their farty methane emissions – not to mention the carbon footprint of transporting meat across the globe and the fuel burned to produce fertiliser – livestock production now accounts for 18 per cent of all greenhouse gases, or 32 million tonnes of carbon dioxide a year, more than all forms of transport, which come in at 13 per cent combined. Still not convinced? Then how about this: the United Nations named cattle as *the* greatest threat to our climate, forests and wildlife.

CHIC FACT: There are over 1.5 billion cattle in the world and a single cow's farts produce enough methane every year to do the same amount of greenhouse damage as four tonnes of CO_2. Methane is twenty times more powerful as a greenhouse gas than CO_2.

Clearly, farming animals for meat is a perilous game, not just for them, but for us and our entire planet. But we promised at the start to explain why meat eating could threaten the *health* of every single one of us, so let's turn our attentions to – the Rise of the Superbugs. (Which isn't, as we would have liked to imagine, some new kids' TV show featuring the superhero cousins of *Mighty Mouse*.) Superbugs are nasty strains of bacteria, like the ghastly MRSA, that are resistant to pretty much every different antibiotic we have in the medicine chest, killing more than 23,000 people every year in the USA alone.

> **GEEK CORNER:** In a big, complex animal like the human being, new generations come along roughly every twenty years or so. In tiny, simple, single-celled organisms like bacteria, new generations come along every twenty minutes or so. Every time genetic material is replicated to create an offspring there's a chance the DNA code won't get copied accurately. Most of the time when this happens it has no impact on how the offspring function, but every now and then a random mutation in the DNA code can give that creature the edge over all the others. Evolution progresses much more quickly in bacteria than humans and with so many animals in which antibiotic resistance *could* potentially develop, it emerges much more often than if they were kept drug free, meaning superbugs are on the rise.

What's that got to do with meat? Apart from guzzling far too many antibiotics ourselves, we are also now feeding them to cattle, chicken and pigs in huge quantities. If you thought that vets use different antibiotics for animals than doctors prescribe for us, you'd be wrong. They use exactly the same ones.

If they had been limited to treating animals only as and when they fall sick with bacteria-related illnesses (antibiotics don't kill viruses) then this superbug predicament might not be such an apocalyptic threat. Unfortunately, in most industrial farming systems, antibiotics are routinely added to animal feed to increase meat yield per gram of food fed to them. An estimated 63,000 tonnes of antibiotics are fed to livestock each and every year, and this looks set to *double* by 2030 if the industry is to keep up with rising demand for meat. The motive for this is purely economic: if antibiotics do the job of keeping bacterial infections at bay, rather than troubling the animal's own immune system with this task, then more energy is saved for other biological tasks such as growing nice and plump and juicy. Plus, fewer animals overall will become ill or die, so from a farmer's point of view it's a win–win situation. In the short term, that is.

The constant presence of these antibiotics in large-scale non-organic animal farming – which has replaced the smaller-scale, less-intensive farming practices of yesteryear – is what is largely driving the development of superbugs. Tiny numbers of bacteria that manage to survive in an animal flooded with antibiotics will quickly multiply in the absence of competition from other bacteria. These will then pass on any chance genetic mutation that accidentally gave them antibiotic resistant qualities in the first place to all subsequent offspring. The result? Millions of new antibiotic-resistant bacteria.

> **GEEK CORNER:** Father of penicillin Louis Pasteur was out collecting fungi in his petri dishes (as you do) and noticed several weeks later that the bacteria that would sometimes accidentally colonise his specimens and were keeping a safe distance from a certain set of his pet moulds. Upon closer inspection he realised they must be kicking out something that killed any bacteria that got close. A number of other scientists were instrumental in finding ways to mass-produce these moulds (in hospital bedpans!), thereby fundamentally changed the outcome of the First World War, but Pasteur is still credited with the initial observation. If he knew what we were doing with the great-great-great antibiotic grandchildren of his wonderful discovery he'd be rolling in his grave.

Once the new superbugs flourish in large enough numbers we may find ourselves in a post-antibiotic era with no effective medicines left to help treat simple bacterial infections in animals or us. Eventually, even the emergency antibiotics that have traditionally been held back from the front line of medicine for use on special occasions (and we're not talking birthdays and bat mitzvahs) will be rendered completely useless. In fact, this has already been documented in parts of India.

Illnesses that are, broadly-speaking, very well controlled at the moment, such as pneumonia and tuberculosis, may become deadly again in as little as a decade or two. Caesareans and other routine surgeries will become too dangerous and without effective antibiotics to give a much-needed helping hand to prop up an already-compromised immune system, chemotherapy may no longer be an option for those battling cancer. We'll effectively have wound the clock back to the bad old days before penicillin revolutionised medicine. And if all this *does* happen, meat eaters – if there are any left – will owe vegetarians one huge, grovelling, post-apocalyptic apology.

Until superbugs get a stranglehold on the human race we are bound to continue to gorge ourselves on meat and food in general. And it turns out that the time of day at which we eat could make a huge difference to how much bigger our eyes are than our bellies.

Late-Night Pickers Wear Big Knickers

It's late at night. You've already had your dinner and you're not really all that hungry, but something is calling to you. Something from deep within the fridge. Something delicious and tasty. It's emitting a low, rumbling sound that you just can't ignore, whispering '*eat me!*' in an alluring, seductive voice. You think to yourself, 'one tiny spoonful can't possibly hurt.' And then, before you know it, you're lying on the floor, spattered head to toe in soft cheese, smothered in cookie crumbs and wondering how it all went so very wrong.

Sound familiar? Probably. Because late-night snacking is a habit many of us indulge in. Unfortunately it turns out this also promotes obesity. Recent research has shown that we are much more likely to pile on the pounds if we eat late in the evening, even if calorie intake stays the same. If only scientists could figure out why we do it then we might be able to start taking steps to curb our enthusiasm for late-night eating.

CHIC FACT: Late-night snacking isn't only bad for your body, it can actually make you more stupid, according to researchers in California. The study only looked at mice, so it's very early days, but it did show that chowing down late at night impacted negatively on their ability to learn.

Well, luckily for us, science is making progress on this front. One study suggested that part of the reason late-night munchies makes us gain weight is that the food we eat in the evening simply doesn't satisfy us as much as the same food eaten earlier in the day. The research found that when people were shown images of high-calorie food, the reward pathways of their brains were not activated as strongly in the evening as they were when shown exactly the same images earlier in the day. The result? We tend to keep shovelling more and more into our mouths in the hope of getting the desired hedonistic food fix at night. While a single slice of cake at teatime can give us the hit we were after, to achieve the same sense of satisfaction late at night we may need to scoff the whole lot to get an equivalent brain high.

CHIC FACT: Humans are born with a genetic predisposition to love sugar, while salt is an acquired taste, one which the snack companies like to tinker with in an effort to make their highly processed, 'nutritionally empty' foods taste more palatable. Some studies have shown that salt is as addictive as cigarettes and that hard drugs trigger a pleasure response in the brain via a similar mechanism. Not all that surprising, then, that companies pack it into their products. The snack-food industry is big business, after all, with big bucks to be made.

Does this mean that it's our brain's fault that we become slaves to the fridge in the wee small hours? Are we helpless to defend ourselves against the power of mind over belly? Not quite. The temptation to gobble down that extra slice of pie during the hours of darkness may not be fully under our control (damn you, brains!), but whether or not we actually act upon this impulse *is* wholly down to us.

The million-dollar question is: what, if anything, can we do to avoid stuffing ourselves silly just before bedtime? Simply being aware that our brains are not as responsive to culinary pleasure late at night is a good start, and you've got that far by reading this book, so give yourself a gold star and a pat on the tummy. Another obvious answer is to remove high-calorie temptations from arm's length and stack the fridge with healthy snacks instead. We are lazy so-and-sos at heart, so if the naughty-but-nice items are still on the shelves of your local convenience store when temptation strikes, the chances of you actually bothering to venture out of the house to buy them are slim. If you're going to gorge more at night at least let it be on something that is good for you.

If all planning fails and you still find yourself tempted by a late-night pigging-out session, just remember – you could always force yourself to go to bed early instead and dream about all the tacos, cheese and ice cream your heart desires.

By this point in the chapter you may be getting a little frustrated that we seem to be telling you that everything is bad for you, but don't worry! We are about to turn the tables with some treats that may actually be good for us. HOORAH!

When 'Bad' Food Comes Good

Why is it that all the most delicious and delightful things in life are also really bad for us? Is it something to do with how we evolved that makes us crave 'naughty' things like sugar, fat and booze which are high in the calories once so essential for our survival? Or is it simply that, by being forbidden fruits (or ice cream, or sweets, or whatever your weakness is), they seem that little bit more irresistible?

Wouldn't it be nice if one day someone turned around to you and said:

> Guess what? You know all those things everyone's been telling you are bad for you? Well, we got it completely wrong. They're actually incredibly good for you. Please, from now on, feel free to subsist entirely on sweets, cheese and champagne.

Sounds too good to be true, doesn't it? But if you spend a little time reading the latest science research, and a few sensationalist newspaper headlines to boot, you'll see that this has actually been happening, albeit in a less-exaggerated form, for some time.

It always puts a greedy smile on our faces when we're told news we really wanted to hear about our favourites guilty pleasures – even if we are well aware that someone else will probably come along to debunk the theory later on. So here are a few of our favourite naughty-foods-we-thought-we-shouldn't-eat-but-could-actually-be-really-good-for-us-and-so-says-science.

Beer

Hops extract has been prescribed in China as a medicine for millennia, although that, of course, doesn't necessarily mean it actually works. After all, certain branches of Chinese medicine also suggest that ground-up rhino horn can cure illnesses such as gout and we know that's a load of old rhino-murdering codswallop (see pp.221–3). Equally, this doesn't necessarily mean the concept that beer comes with health benefits is completely unfounded. In fact, happily for the pint swillers among us, modern scientific testing has suggested that the beery blessing does indeed boast many impressive qualities. The active ingredient of hops extract – xanthohumol – is known to kill viruses, reduce inflammation, protect our brains from degeneration, fight prostate cancer, protect the cardiovascular system and even reduce obesity.

Hang on a second: did you say *reduce* obesity?

> **GEEK CORNER:** A Dutch study in 2000 showed that men who drank beer had a 30 per cent increase in vitamin B6 in the blood plasma after three weeks. B6 prevents the build-up of homocysteine in the body, which has been linked to heart disease.

CHIC FACT: Traditionally, Europeans have always drunk a lot of beer. Although these days it is often associated with loutish behaviour, we may actually have it to thank for the survival of our ancestors. Brewing weakly alcoholic ales, meads and ciders meant the toxic action of the alcohol on micro-organisms in the water made it safe to drink. Beer, with its extra payload of xanthohumol-related health benefits, has therefore helped sustain our ancestors for thousands of years.

Beer guts up and down the country may well be lining up to scream, 'that can't possibly be true!' and it's a fair point because, while xanthohumol may well push back against obesity, there are also an awful lot of calorie-laden carbohydrates in beer too. The Czechs even acknowledge this with their nickname for the amber nectar, which translates as 'liquid bread'. The more beer you drink, the more sugar juice you're consuming, which eventually gets packed away into fat stores under the skin, hence the beer belly. However, don't lose hope; there is plenty of data to indicate that overall, if enjoyed in moderation, regular beer drinking can be good for your heart and increase your lifespan. Whether this is down to xanthohumol, or the increase in vitamin B6 seen in beer drinkers' blood, or something else altogether we don't know for sure, but a correlation has been observed.

All hail the mighty pint!

> **GEEK CORNER:** At very high doses of purified xantho-humol extract – equivalent to around 2,000 litres' worth of beer per day – this wonder ingredient really starts to work its magic. Researchers in Langzhou, China, recently published a paper suggesting that the antioxidant activity of xantho-humol could protect against Alzheimer's and Parkinson's diseases.

Junk Food

You may think we're now taking things too far now. Anyone who's ever watched the American documentary *Super Size Me*, picked up a women's mag, or, you know, spoken to another human being, knows that too much fast food is a sure-fire path to obesity. So imagine our shock when newspaper headlines in 2015 implied that fast food could actually carry some health benefits. This was based on the results of a study that suggested eating fast food after a workout could help you recover more quickly and even build muscle. A headline-grabbing and eyebrow-raising claim indeed! So, like the super-sleuths we are at heart, we set out to investigate.

After some digging around, we found that the research was actually very specific, comparing how various foods enhance glycogen recovery after exercise. After a hard-core, intensive workout – a concept that one of your authors still struggles to grapple with – your body cries out for nutrients to refuel your muscles' primary source of stored energy: glycogen. Without replacing the glycogen stores, our metabolic system looks elsewhere for sources of energy, weakening the body and making training later on less effective. Athletes on intensive training schedules need to be able to exercise more than once a day, so how can they do this without weakening their muscles?

Step forward heavily marketed, pricey sports supplements, drinks, nutrition bars, gels and powders that promise to replenish glycogen

stores and get you match-fit for your next workout. Happy days, indeed, but they come at a cost: post-workout supplements for those pumping iron are a multimillion pound industry. Seeing this, one group of scientists began to wonder whether it was really worth the price tag. Could gobbling down Maccy D's finest do the job just as well?

Participants in the study all worked out for a gruelling, sweaty ninety minutes, pretty much wiping out their glycogen stores. This was followed by a period of 'recovery' in which one group ate Ronald McDonald's dream dinner: hotcakes, hash browns and orange juice, followed two hours later by burgers, fries and Cokes. The other group had to make do with the traditional athlete's fare of sports' drinks and power bars also eaten at two hour intervals. Amazingly, they found that no matter what the athletes ate, their glycogen levels rose to about the same level. Both groups were just as ready for their next hardcore workout.

So what exactly does this mean? Can we hit the treadmill at Barry's Bootcamp with a cheeseburger waiting for us in the changing room? Can we chuck out the NutriBullet and bring back the deep-fat fryer?

Unsurprisingly, the answer is a resounding 'no'. As we mentioned, the study was limited to one small area alone: the ability of various foodstuffs to replenish glycogen stores after a workout to get you ready for another session soon after. And, OK, this ability to exercise many times a day *is* an important and key part of many athletes' intensive regimes but it isn't the only factor they, or we, need to consider. Those delicious vessels of naughtiness – salt, cholesterol and sugar – found in high levels in junk food have been associated with all sorts of health problems and certainly won't do any favours for someone trying to hit peak physical condition. And, let's be honest, while the Olympian specimens among us may need to work out more than once a day, how often do we mere mortals ever really exercise just a few hours after our last exhausting session? It does give you pause for thought, though: when feeling low on energy, or looking to recover quickly after exercise, do we really need to shell out wads of cash on special sports supplements – or could a healthy snack like fruit or wholegrain carbs do the trick just as well?

Chocolate

Chocolate is one of the most lusted-after foods in the world. A delicious, comforting, tasty, joy of a treat. Kit Kats, Terry's Chocolate Orange, Twix, Mars Bars – whatever your choc of choice, almost everyone adores it. An estimated one billion people (that's one in seven!) chow down on its velvety goodness every single day; nine out of ten people say they 'love it' and half the population (who definitely aren't prone to exaggeration) say that they 'can't live without it'. Yet the sugar and cream pumped into modern-day chocolate make it a guilty pleasure that many of feel we need to cut down on. Wouldn't it be nice if it was actually good for us?

1. *Good for your heart*: Several studies have shown that dark chocolate can have serious benefits for your heart. In 2014, researchers found that dark chocolate makes arteries in the heart more flexible and stops white blood cells from sticking to the blood vessel walls, preventing arteries from getting clogged up by fatty substances, a condition known as atherosclerosis. Other studies went on to show that chocolate can lower blood pressure and reduce the risk of cardiovascular disease by increasing levels of 'good cholesterol' (high-density lipoproteins) and lowering levels of 'bad cholesterol' (low-density lipoproteins), the combination of which helps to stop all that gunk getting lodged in our arteries. And in 2011 the news we had all been waiting for was revealed when researchers found that people who ate dark chocolate more than five times a week had a whopping 57 per cent less chance of coronary heart disease. Hooray for the dark stuff!

2. *Good for your skin*: That's right, scoffing chocolate as you dream about getting some holiday sun isn't a delicious fantasy; it could also be a wise move if you want good

skin. The flavonols in dark chocolate can improve hydration, protect your skin from sun damage and generally improve its condition and appearance.

3. *Good for your brain*: It's a sad fact of life that as we get older our ability to process and remember information starts to decline naturally but – good news just in – a 2016 study showed that regularly eating chocolate long term could prevent age-related cognitive decline. The research studied the habits of 1,000 people and discovered that those who ate chocolate more often also performed better on a whole range of memory and reasoning brain tasks. It's likely that natural chocolate polyphenol compounds called flavonoids are key to these results, with other research showing that eating these can improve cognitive function in the short term. Suddenly the tagline 'only Smarties have the answer' makes a whole lot more sense.

All those health benefits are just the tip of the choc-iceberg. Other studies have claimed that dark chocolate can do everything from protecting against strokes, increasing libido (see pp.166–70), decreasing chances of diabetes, lowering blood pressure, reducing stress and even helping you lose weight. The perfect excuse to shovel it into our gobs!

Perhaps the Aztecs and Mayans had it right when they said that chocolate was the food of the gods . . .

Delighted as we are that chocolate can be good for our health we do have a quick caveat – dark chocolate is best, as the added sugar and cream in milk chocolate carries other health risks (sorry Milky Way, but you're going in the bin). However, recent scientific research has found new production methods that make chocolate naturally sweeter, meaning less sugar added to it. And for that feat we think science deserves a little thank-you note.

A Letter of Thanks to Science for Chocolate

Dear Science,

You beautiful, beautiful thing. Over the years you've done so much incredible stuff; you've discovered vaccines, you've

saved lives and you've created all sorts of new life-enhancing innovations, but now Science, *now*, you have truly outdone yourself. You've not only gone and found a way to make chocolate taste more delicious – which is a feat worthy of our love and adulation all by itself – but you've also made it healthier at the same time. Bravo!

By tweaking just a few things in the early and delightful process that is 'making chocolate', you've been able to boost the levels of those magical health-giving properties found in cocoa beans called polyphenols, and in particular flavonoids. These plant compounds – so rich in anti-inflammatory and antioxidant properties – have been known to improve our health, lower our blood pressure and reduce our chance of cancer, stroke and heart disease. But traditional chocolate-making processes, like drying, roasting and fermenting the beans, cruelly destroy so many of these health-giving polyphenols. If only there was a way to create chocolate without destroying so much of the good stuff . . .

Oh Science, we jest, because we know only too well that this is precisely what you've done. Armed with your team of researchers at the University of Ghana, you figured out that by storing the pods for seven days before the fermentation process, and by roasting the beans for a longer time but at a lower temperature, you could preserve the greatest amount of polyphenols with the highest levels of antioxidants.

Not only that, Science, but this new and improved process has a beautiful and surprising side effect. It actually makes the chocolate taste sweeter! By keeping the cocoa beans in their pod for that extra week they have longer to soak in the sweetness of the surrounding pulp. The final chocolate product needs much less of the substance we hate to love and love to hate: sugar. Bravo again. You have truly excelled yourself.

Do you even know how long we humans have been munching chocolate and cocoa? The ancient Mayan people saw cocoa as a symbol of life itself – the food of the gods. The Aztecs believed it gave them power and wisdom and used it as a currency – yes, chocolate money was an actual thing. Both Aztecs and Mayans made the earliest hot chocolate, mixing cocoa with hot water and some chilli peppers for spice.

Then, in 1528, Hernán Cortés brought cocoa to Europe, adding sugar, vanilla and all sorts of tasty spices, elevating the humble cocoa bean to an indulgence of the very first order.

And on one glorious day in the eighteenth century, chocolate bars were finally produced en masse, with factories squeezing cocoa butter from the cocoa beans in huge quantities. Delicious.

Since then, we who worship at the altar of chocolate have been gobbling it down like there's no tomorrow, all the time believing that it was really bad for us. Well, no more, Science, no more! You have changed everything.

Science, we know it is early days and there is so much more work to be done, but for discovering this new technique of making delicious dark chocolate which is healthier and naturally sweeter and therefore almost guilt free, we will remain for ever in your debt.

With love, respect and gratitude,
Chocoholics of the World.

While we're busy thanking science for giving us the gift of guilt-free cocoa, another tasty bean has been quietly working miracles for us over the centuries, too.

Coffee Brain

Look back at newspaper headlines about coffee over the last 500 years and you'll see one of the finest examples of mixed messages in the media ever. In the sixteenth century it was blamed for promoting illegal sex, in the seventeenth century it became considered as something of a cure-all, in the eighteenth century it was praised for helping people increase their work productivity, and by the nineteenth century it was being blamed for making people go blind. But we still weren't done with poor old coffee bean. In the early twentieth century it was being blamed for stunting people's growth and sending children's school grades tumbling, only to be cited as a cause of serious health problems like heart attacks a few decades later as the twenty-first century loomed on the horizon. Then, at the turn of the millennium, everything changed. Clarity was finally carved out of the confusion and today coffee has finally been placed in a seemingly universally positive light – when consumed correctly, it's something of a wonder bean.

> **CHIC FACT:** A dose of 10 grams of caffeine is deadly. Given that a typical cup of brewed coffee contains 100 milligrams of caffeine you'd be well advised to step away from the roaster if you find yourself approaching 100 cups in a day, or you could find yourself popping your clogs.

You see, the caffeine in coffee does brilliant things for our bodies and brains, keeping our minds alert and even warding off diseases. It works primarily by blocking the receptors of an inhibitory neurotransmitter called adenosine. Under normal circumstances, the job of this chemical messenger is to dampen the excitability of wire-like neurons that ferry billions of electrical messages that criss-cross the whole brain every second of every day. By blocking these receptors and removing the inhibitory influence on brain activity, caffeine actually *increases* activity in the pathways involved in alertness/attention/focus as well as those that initiate body movements, which is why too much can give us the shakes.

Coffee comes in handy for boosting concentration levels, but only up to a point. There's a sweet spot in which you feel more alert and switched on, but beyond that you can become so keyed up that you start to get distracted and flustered. And, trust us, caffeinism* is not a good look.

As for long-term brain benefits, there are plenty. Regular drinkers of moderate amounts of caffeine (three to five cups a day) have been found to have a lower incidence of Parkinson's, Alzheimer's, liver and heart diseases. Other studies have observed that it seems to protect against stroke and certain forms of cancer. The jury's still out on the precise mechanism at play for these health benefits, but the evidence that moderate amounts of caffeine have a neuroprotective influence on the brain is steadily increasing. Happy days for coffee addicts.

CHIC FACT: Caffeine is found in several plants other than coffee, such as the kola nut (one of the original ingredients of Coca-Cola) and guarana, a wonder berry from the Brazilian rainforest. It's also found in low quantities in chocolate. Caffeine is also included as a stimulant in many cold and flu remedies – so beware what you reach for when you wake up in the middle of the night with a bunged-up nose!

Before you run off to spray half a dozen Americanos down your throat, a little housekeeping. If you want to get all the benefits of coffee whilst simultaneously avoiding the many pitfalls, there are a couple of things you need to be aware of. Caffeine is the arch enemy of sleep and getting enough good-quality zeds might be the best possible thing you could do to help keep your brain healthy. You don't want to be drinking your caffeine late in the day, despite what every

---

* A very high but not deadly dose of coffee can lead to a quite severe psychiatric condition known as caffeinism, characterised by restlessness, agitation, excitement, rambling thought and speech and insomnia.

waiter in the Western world would have you believe when they offer
you a cheeky after-dinner espresso.

CHIC FACT: We drink around seventy million cups of coffee
every day in the UK. That's an awful lot of coffee beans – it takes
forty-two beans to make an espresso.

When exactly should you drink it? The key to figuring this out
is knowing what the half-life of caffeine is (you may have only
previously heard the term 'half-life' in the context of radioactive
substances or zombie video games, but it's a really important concept
to wrap your grey matter around if you want to enjoy the benefits of
coffee without screwing up your ability to get a good night's sleep).
Caffeine's half-life is six hours, meaning it takes six whole hours for
the concentration of caffeine in your bloodstream to halve. Imagine
you've had four to five cups of coffee – each containing 100 milligrams of caffeine – over the course of the morning. Let's say then
that there's 400 milligrams of caffeine in your blood at midday and
you've got a bit of a buzz on. All that caffeine is still going to stick
around in your system for ages. Even if you don't drink a single drop
of caffeine for the whole rest of the day, at 6 p.m. the 400 milligrams
will only have reduced by half to 200 milligrams and another six
hours later at midnight it will have halved again, down to 100 milligrams. In other words, come midnight, when your brain wants to
sleep, you'll still have a full cup of coffee's worth of caffeine swimming around in your system. Our advice? Consume your caffeine
early in the day, avoiding afternoon and evening coffees like the
plague. That way enough of it will have been removed from your
system by the time you hit the sack and so you'll be able to get a really
good night's sleep.

There are many more pros and cons to the nation's favourite
drinkable bean. But, despite the drawbacks, when drunk strategically
(i.e. early in the day) and in moderation, a few cups – no more than
five! – of the liquid black gold can not only be delicious, but should
help keep the doctor away, too. It is the closest thing we have to a
natural genuine smart drug.

GEEK CORNER: The half-life of caffeine also depends on what other drugs you're taking. It takes up to twice as long for the liver to remove caffeine if a woman is on the contraceptive pill. In other words, the half-life increases to twelve hours instead of the usual six hours. Conversely, in a smoker, caffeine is removed much faster than in a non-smoker. The half-life for caffeine in a smoker's body is just three hours.

You may have been surprised to learn that a morning coffee habit could be associated with so many health benefits. But what about our other favourite liquid indulgence? Could a glass of wine be the perfect excuse you've been looking for to skip the gym?

Wine Versus the Gym

Picture this. You arrive home exhausted after a long day's work and the doorbell rings. You answer it and your heart sinks as you realise it's your friend who you said you'd go training with. In they saunter and sit down while you set off in search of your trainers, resigning yourself to 'no pain, no gain'. Your friend halts you in your tracks, grinning widely, and holds up a bottle of vino, saying: 'Leave the trainers – this is tonight's workout and it's going to be just as good for us.'

Sounds too good to be true, doesn't it? But not according to some headlines in 2015, when the media got into a big old frenzy about a study on the effects of an antioxidant found in red wine, and declared loudly and proudly that a glass of wine was 'as good for you as an hour in the gym'. Lushes around the country rejoiced, while exercise bunnies sighed knowingly. Dedicated science afficionados that we are, we immediately set to work to verify these claims.

We found that the study in question was indeed pretty groundbreaking. It showed that resveratrol – an antioxidant found in red wine as well as in blueberries, raspberries, mulberries and some nuts – *does* come with certain health benefits that mimic some of the many positive effects of exercise. A high dose of resveratrol – a polyphenol compound produced naturally by some plants in response to injury or bacterial attack – improved participants' heart function, physical performance, metabolism and muscle strength in a similar way to an extensive training session. Hoorah! Case closed. Let's all skip the gym and grab a bottle!

Not quite so fast, because here comes the rather big 'but' – the patients tested in this study were in fact ... rats. Most newspapers didn't seem bothered by this small but important fact when they wrote up their articles. Add to this the fact that the rats weren't fed the resveratrol in tiny little mouse-sized glasses of wine – they were instead fed pure resveratrol without all the other booze bits – and you begin to see that some in the media may have jumped to a conclusion or five.

In fact, the resveratrol, already known to have benefits for our hearts and muscles, was also given in conjunction with exercise. The rats all took part in a twelve-week endurance training programme, running on miniature treadmills at regular intervals. Those also fed a diet of resveratrol showed better endurance, increased muscle strength, boosted metabolism and heart function. Effectively, what the study showed was that resveratrol is a performance booster, which could eventually come in handy for athletes hoping to get the most from their training, or for people with conditions that prevent them from exercising.

People with type 2 diabetes, for example, who are at greater risk of heart failure during bouts of exercise, could benefit from it, a theory that the same team are currently putting to the test. The actual human beings taking part in this new study are given around 200 milligrams of resveratrol a day and to get the same amount from the old plonk you'd have to guzzle somewhere between 100 and 1,000 bottles of wine a day.

Have we crushed all the wine-drinking joy out of your lives? Hopefully not, as the overall benefits of red wine in moderation are well documented, including fighting cancer and lowering risk of stroke – so having a little red wine each day as part of a healthy diet could well be a positive thing.

The next time your friend pops over to entice you to go for an invigorating run say yes, you'll go, but afterwards you'd like to sit down and shoot the breeze over a cheeky glass of resveratrol-packed red wine for good measure.

While some food or drink we traditionally thought of as bad for us may in fact carry some health benefits — like wine, coffee and chocolate — it's still more important than ever to eat traditionally healthy foods, too — like fruit, vegetables, wholegrains and oily fish. For those of us who struggle to force ourselves to eat and enjoy such healthy foods there is a tasty light at the end of the tunnel: we may be able to retrain our brains to become 'addicted' to healthy food.

Healthy Food Addiction

If someone leaned over to you at a dinner party and promptly announced, 'oh darling, it's just so tough for me, I'm *soooo* addicted to healthy food', you might be forgiven for wanting to poke them in their smug little faces. However, if they told you that you too could share in their addiction with a few simple techniques you might find yourself genuinely intrigued instead.

The trouble with sweet and fatty foods is that they trigger a huge response in the brain's reward pathways, which regulate how much pleasure we feel, not to mention strongly influencing future choices. This harks back to our ancestors' cavemen and cavewomen days when those who instinctively favoured high-calorie food options were much less likely to starve, so these are the calorie-loving brains we've inherited — cheers, evolution!

We all have this natural inbuilt preference for a bit of the old naughty-but-nice. But in this food-saturated modern world of ours, if these impulses aren't curbed then eventually we could become obese or have high cholesterol. And the worry with obesity isn't so much to do with the fat that changes our body shape in a way that's visible from the outside. It's more to do with an excess of fats and lipids floating around in our bodies which packs itself in around our body's internal organs and also meddles with certain important brain cells in the hypothalamus. This is why being thin with high cholesterol is just as bad as being obese. The hypothalamus sits at the base of the brain and is effectively the Master of Hormones (a brilliant new name for a superhero perhaps?), regulating all sorts of bodily systems. The brains of some obese people show inflammation in the parts of the hypothalamus that oversee appetite regulation, which may well mean that the more unhealthy food they eat the less they can control their eating compulsions. This vicious cycle can be a slippery slope and it's not only bad for the people who are doing the munching.

The numbers of obese people in the world are rising at such an alarming rate that obesity-related illnesses are now placing immense pressure on global healthcare systems. A scientific breakthrough like the one we're about to relate might save lives.

It's all about retraining people's brains to feel more pleasure when they chomp down on healthy foods, which eventually makes them crave healthy food instead.*

The simple solution to defusing the ticking time bomb of obesity was uncovered during a small pilot study called iDiet. Thirteen overweight and obese punters had fMRI brain scans before and six months after their participation in the programme. During that time the dieters ate low GI carbs, high fibre and protein like fresh veg and wholegrains, and that daily calorie intake was reduced to 1,000.

CHIC FACT: The term 'junk food' was first coined in the 1960s but only became popular after 1976 when the 'Junk Food Junkie' song topped the charts.

Not only did the iDieters lose weight, an average of 6.5 kilograms, but the fMRI scans revealed that their brains had actually been retrained! After the trial their brains showed a greater pleasure response to the healthy low GI food *and* a smaller reward response to the high GI stuff like cakes, booze and biscuits. In other words, they

---

* By healthy foods we mean 'low glycaemic index' (low GI), fodder like fresh fruit, vegetables and wholegrains that release sugars into the bloodstream slowly, a bit like a drip feed, rather than dumping it in all in one go. Unhealthy foods are those tempting fast-release carbohydrates, i.e. with a 'high glycaemic index' (high GI) – foods like bread, soft drinks, cakes, booze, biscuits, pizzas, ketchup-laden chips and so on – which dump sugars straight into the bloodstream, causing glucose levels to shoot up to levels that are potentially very dangerous for body and brain. If you are not being particularly physically active in the near future then you'll never manage to use it all up and so all the excess sugar gets converted into fat; stored around your organs most lethally, but also bulging beneath your clothes. You'll also get hungry again sooner, damn it!

were now getting more genuine enjoyment out of eating healthy food as a result of the intervention and less pleasure from the bad stuff.

It looks like the vicious cycle that can lead to obesity can also work the other way round. The more healthy the scran you put in your belly when you're hungry, the more pleasure you'll end up getting from it. It will soon become much easier to stick to the good stuff, and then one day, who knows, you may find yourself sitting at home moaning and groaning with pleasure over a stick of celery.*

It's incredible to consider the advances in science that are challenging the way we think about food and drink. They can make chocolate healthier, teach us how to become addicted to healthy food and even, it seems, turn food into jewels.

Turning Peanut Butter into Diamonds

Diamonds may well be a girl's best friend but so, too, it seems, is peanut butter. At least, that's what an experiment by a group of scientists in Germanys seem to suggest, after it managed to turn a blob of the UK's third best-selling breakfast spread into our most beloved gemstone.

> **CHIC FACT:** We have De Beers to thank for the expression 'diamonds are a girl's best friend'. They planted the idea that the only way to make a truly romantic gesture was to gift one of these hugely expensive sparkly rocks as proof of true love in adverts starting in the 1940s. Marilyn Monroe certainly took their word for it.

The key to the experiment was that, like all foodstuffs – and all living things, for that matter – peanut butter is made up of carbon-based compounds. Diamonds are made up of carbon, too, just a whole lot denser than in its peanutty form and packed into a rigid crystal lattice structure. With a clever bit of jiggery-pokery the German team was able reconfigure the molecular structure of the carbon in the

---

* OK, this may be a slight exaggeration, but you will definitely get a greater pleasure response in your brain to it than before. Pass the hummus!

peanuts and turn it into more far more valuable bejewelled cousin.

To do so, they had to recreate the conditions found in the Earth's lower mantle (550 miles beneath the surface) where diamonds are naturally formed. A combination of piston presses, anvils and burning hot furnaces was used to reach temperatures of 2,000°C and pressures 1.3 million times that of the air that we breathe.

The peanut butter was subjected to all this extreme heat and pressure for a period of about two weeks using a process known as the 'stiletto heel' – an anvil, made of two tiny diamonds itself, used to squeeze the tiny blob of peanut butter between them, as the furnace heated it up. The carbon within was slowly but surely converted into a denser crystallised form and, hey presto, a tiny diamond was born.

Have we been spreading a potential fortune on our toast and gobbling it down without so much as a thought? Not really. Because before you run out to the shops and empty the shelves of the entire stock of Sun-Pat, it's important to remember that the process is difficult, very expensive and incredibly slow. The diamonds took weeks to form and used up huge amounts of energy. Besides, because of impurities in the peanut butter, the resulting precious stones were muddy looking, of low quality and pretty tiny to boot, only about two millimetres across.

Why exactly did they decide to do it? It certainly wasn't to give London's Hatton Garden jewellers a run for their money. Lead scientist Dan Frost and his team happened to have been working for some time on trying to recreate the conditions of the Earth's core, in order to deepen our understanding of how our planet was first created. As part of this research they had already starting turning rocks into diamonds, and hearing about his work, a German TV company challenged him to the peanutty feat, and who can resist a dare?

You'd be forgiven for wondering if there's actually any use for the peanut-converted diamonds, other than making us feel excited that we own a tub of potential diamonds in our cupboards. Happily there is – as well as being used in machines for grinding and cutting, synthetic diamonds like these are used in the very devices (remember the diamonds in the 'stiletto heel'?) that produce the astronomically high pressures needed to work out how the Earth was formed in the first place. Not bad!

Now, all we need is for scientists to figure out how to turn Marmite into Ferraris and we're set.

GEEK CORNER: Any foodstuff or carbon-rich object could be turned into a diamond under the same conditions to which the peanut butter was subjected. The only problem is that this process creates a lot of hydrogen as a waste product, which has a tendency to explode and damage the physicists' precious equipment!

The big question now is how do we end a chapter about the wondrous innovations and latest discoveries about food? With a story about our food-loving chimp cousins, of course!

Chimp Chefs

We all know that chimps like a tea party. Or at least that's what the controversial PG Tips television ads of the past had us believe. But can you really imagine a chimp chef baking cakes and brewing tea, let alone prepping veg, searing steaks or baking pies? Sure, they'd look swish in their chimpy chef's hats and aprons, but the moment a sharp knife appears in an ape's hand, you'd be right to feel concerned.

That said, it does seem that if chimpanzees *could* cook, they would. They might lack the manual dexterity, fire-making and planning skills required to prepare a meal, but they will happily invest a bit of time and effort if it means getting to eat cooked rather than raw food.

How in the name of Tarzan do we know this?

GEEK CORNER: Chimpanzees learn from each other. Usually chimps use crunched-up leaves to soak up water from tree trunk hollows or crevices in the jungle floor. However, one individual under observation spontaneously worked out that ripping up moss and using it as a sponge was a more effective tool. Subsequently, chimpanzees that witnessed this approach quickly adopted the technique, even passing it on to others in the group. Cultural transmission in the chimpanzee. Awesome.

A recent study by Harvard scientists offered a bunch of chimps a menu choice, along the lines of 'what would Sir prefer tonight, the cooked parsnip or the salad?' OK, not *quite* like that, but they did find that the chimps generally chose to pick up and eat cooked food over raw when both were put in their enclosure.

A cunning study that took place deep in the Congolese jungle then took the investigation one important step further. The researchers wanted to know whether, given the right cooking tools, primates would use them to cook their food, even if that meant waiting a bit longer to chow down. Having the willpower to wait patiently for food in order for it to taste better when it cooks isn't easy – it's something even *we* struggle with daily. How many times have you found yourself impatiently gobbling down cheese on toast before it's all bubbly and delicious? Or pushed the concept of al dente pasta to its absolute max?

The research team used a 'magical cooking machine' to convince our *pan troglodyte* cousins that if they put raw food in at the top, after a short wait it would reappear cooked to perfection at the bottom. Unbeknownst to the chimps the researchers were really just removing raw materials from the top section and placing the fully cooked meal into the compartment at the bottom. Once they'd learned how to use the system, 90 per cent of them chose to 'cook' their food, popping the raw fodder in up top and retrieving the tasty cooked food from below, even when that meant that had to wait for up to a minute. Their ability to resist their natural urge to wolf down the raw food immediately* demonstrates a surprising degree of self-control in waiting for the food to cook *and* an impressive ability to plan for the future.

CHIC FACT: Chimps and people aren't the only ones who favour cooked food. Cats prefer cooked meat and rats go wild for cooked starch.

---

* Chimps are thought to have evolved with the desire to eat food straight away as they are usually found in groups, or whoops, so competition over food is fierce, and the only really safe place for it is tucked away in your belly, away from thieving hands.

These findings are all the more remarkable when you bear in mind that part of the study actually required the chimps to carry the food a short distance to get it to the oven. As knuckle-walkers they had to carry the raw food in their mouths during the journey, making it even harder to resist the urge to take a cheeky bite. We're not even sure we could hold out against such temptation – any morsel of food that ends up finding its way within five inches of our mouths usually gets devoured within seconds. So the fact that 60 per cent of the chimps resisted doing the same thing showed how much they prefer cooked food, and that, given the right incentives, they are able to inhibit their strong natural impulses. Of course, we must remember that the fact remains that chimps still don't have the skills needed to *actually* cook for themselves; however, the desire is there, and if they could, they would.

This chimp study also revealed a little something about our own species. Demonstrating that chimps are prepared to do cooking-related work suggests that they share our own preference for cooked foods over raw, which makes it much more likely that the common ancestor of the chimp and human that lived about six to seven million years ago most likely felt the same way. Learning to cook probably helped us to evolve into our present form by giving us access to nutrients that can't be absorbed when food is eaten raw. The main difference in our evolutionary outcomes is that we humans learned to control fire to cook foods in order to gain access to those all-important brain-building nutrients and the chimp side of the family tree did not.

Perhaps most importantly of all, however, is that we suddenly understand why King Louie in *The Jungle Book* was so desperate to get Mowgli to teach him the gift of fire – the royal orang-utan probably just wanted to be able to cook himself a nice tasty loaf of banana bread. Can you blame him?

Final Thoughts

You'd be forgiven for feeling a little topsy-turvy about food after reading this chapter. Are fish really on drugs? Is cheese really like smack? Does Gordon Ramsay have to watch his back cos there's a new chimp in town?

Our suggestion to you is to take all these stories with a tiny pinch of salt (and pepper). New research may well soon emerge to contradict much of what we've shared with you here, and then re-prove it a little while later.

Eating a varied, balanced diet of what you already know to be good, nutritious wholefoods, without denying yourself some of the naughty stuff on occasion, is always a good policy. Food can and should be a great source of pleasure, and sitting down and eating a meal together will always be one of the very best ways to bond socially with our fellow human beings. Besides, food is the very thing that connects us humans to our planet; just like fruit and veg, our own bodies are constructed from molecules that have come out of the Earth itself. It may seem obvious, but it's still pretty incredible.

As for the apes, we're pretty sure that if chimps did ever learn to cook by themselves then dining in a jungle restaurant boasting a chimp head chef could be fun. We'd certainly enjoy eating banana-shaped spaghetti hoops.

Well, here we are. One chapter left, and we promise it's going to be a good one. Maybe science is the answer to vanquishing the bad guys and saving the world . . . pull on your Lycra onesies folks, and make sure you're wearing your pants over your leggings, the superheroes are coming!

9

We Could Be Heroes (Just for One Day)

We human beings have never been satisfied with our everyday, boring old conventional capabilities, have we? We sing about, write about, talk about, dress up as and dream about superheroes from the earliest of ages. It may be that you spent endless childhood hours pretending to be Superman, ThunderCats or Spider-Man (or Batfink in the case of Lliana and Dr Jack). Or as an adult you may enjoy losing yourself for hours in films like *Birdman* and *X-Men* or, better yet, heading out to fancy dress parties dressed up as Mighty Mouse, Wonder Woman or Batman.

Being a superhero is the ultimate fantasy for many of us and having superpowers is half the battle towards achieving this lofty dream. Let's imagine for one moment that we can suspend the dull constraints of everyday life and believe, in the immortal words of the late, great David Bowie that, 'we could be heroes, just for one day'. What powers would you wish for? And, don't worry, this isn't like the genie in the lamp; we're not limiting you to a measly three wishes. You can have as many as you want, so go ahead and be greedy. Take the whole kit and caboodle, should you desire.

Superhuman strength? Sure. Invisibility? Why not? To soar through the air? Knock yourself out! On second thoughts, maybe flying is a bit slow for our needs. After all, if we've only got twenty-four hours in which to enjoy our various incredible new abilities we'll want to pack as much in as possible so maybe we should go for teleportation instead.

Although, if we'll be quantum leaping from place to place we're really going to need Neo's *Matrix*-style capacity for downloading new skills directly into our brains. Speaking fluently in all languages, for example, would help us get so much more out of our travels. And being able to breathe underwater is also a must to take advantage of all the fun that's to be had under the sea.

Clearly this day of superheroic madness won't all be plain sailing either; for every Superman there's some kryptonite, for every Bananaman there's a Doctor Gloom. So we'd also need to take steps to mitigate against various possible disasters we might face along the way. Being able to survive extreme conditions and the ability to self-heal might both be prudent measures.

Think we're just teasing you with all this? That we're cruelly dangling a superhero carrot in front you with no hope of you ever getting a juicy bite? We wouldn't do that to you, dear friends. The truth is that science is an incredible thing (if you hadn't noticed yet, we're kind of big fans), constantly developing new innovations and technologies, pushing us beyond the limits of our own capabilities, opening up an Aladdin's cave of incredible superhero-esque possibilities.

We're going to guide you through the latest, weirdest, craziest scientific advances that might bring some of those seemingly impossible superpowers within reach. Some require hard work and dedication, others are simply a matter of smart tech, but all of them take us a step closer to realising our childhood superhero dreams, so first up, let's look at some specific superhero role models:

Superman – How to Fly

He's the quintessential superhero. The puff-chested, chisel-jawed, red-booted, world-saving, cape-wearing, transmogrifying Man of Steel we love to love. But the most iconic part of Superman's persona – apart from the super-speedy telephone booth costume changes from journalistic geek to superhero chic – is the fact that he can fly.

In the comics, originally created by artist Joe Shuster and writer Jerry Siegel way back in 1933, Krypton-born Superman is able to fly because, compared to the planet Krypton, Earth has much less gravitational pull holding our hero down. His body is also solar-powered *and* he can produce his own gravitational force that he can point any-which-way he pleases. How plausible . . .

CHIC FACT: Krypton was red when it first appeared in the Superman comics, although it was later portrayed as the recognisable green. Other variations have also appeared in the comics: gold kryptonite (which strips Kryptonians of their power permanently), white kryptonite (which kills plant life) and pink kryptonite (which made Superman more effeminate, exclaiming things such 'did I ever tell you how smashing you look in bow ties?'). However, our all-time favourite was the periwinkle kryptonite, which makes Superman hallucinate psychedelic disco scenes and dance about.

Lucky Clark Kent, but what about the rest of us mere Earth-born mortals? The ones who dream (literally in our cases) of flying around exploring the world, reaching dizzying heights, or simply pulling off loop-the-loops in our back gardens. Helicopters, aeroplanes, hang-gliding, bungee jumping and floating in space (à la astronautical adventures) are just a few of the things some of us have already been lucky enough to try. But as time and technology move forward our options are now getting much closer to enabling us to realise our bird-like desires. So, how do they measure up to the real Superman?

Troy Hartman, the real-life jetpack man, uses a rather cumbersome set-up including a jet-thruster backpack in combo with a kite surf-style kite to keep him airborne. He does manage to successfully propel himself through the air but, truth be told, pretty slowly, and the kite leaves him very prone to being buffeted around by the wind. Superhero FAIL. Sorry, Troy.

Then there's Glenn Martin, who, in 1981, was struck by a bold dream. Thirty years of research later – including much tinkering in his garage, searching for funding, experimentation and endless rounds of refinement – and the fan-driven Martin Jetpack finally flew thousands of feet above terra firma. A major hurdle involved proving that the specially designed rocket-deployed parachute for emergency landings could deliver both jetpack and crash-test dummy back down to Earth safely. Ultimately he's targeting sales at various institutions involved in search-and-rescue missions, so don't be surprised if you start seeing them up in the skies over the next decade or so. This jetpack is truly awesome and you can even try it out in virtual reality before you buy. The only catch is that the flying posture is very, well, vertical, and we want to fly like Superman. So we're really sorry, Glen, it's a case of close, but no cigar.

> **CHIC FACT:** The creators of Superman, high-school students Jerry Siegel and Joe Shuster, sold the rights to Superman for just $120. Talk about getting royally ripped off! The pair never knew how successful their creation would be and never managed to recover the money they felt they were owed for it after their idea hit the big time.

The 2016 Consumer Electronics Show in Las Vegas also looked promising with the unveiling of Chinese firm Ehang's quadcopter for human flight. In test flights it's been whizzing around the Guangzhou countryside at 60 mph, which would be great for commuting but is hardly the stuff of superheroes. It looks like a drone quadcopter, but scaled up to accommodate a human-sized passenger in its sports caresque driving compartment. An important further addition includes double propellers at each of its four corners to increase thrust. So it looks good and certainly brings us one step closer to George Orwell's vision in his book *1984* of everyday people zipping about overhead on their own little personal flying machines.* But it's basically a helicopter with four sets of propellers instead of two. Superhero FAIL again, we're sorry to say.

As you can see, all of these options fall short of the real Superman deal. None of them allow us to propel ourselves through the air, fist thrust out before us, changing direction at will and reaching superspeeds – but, fear not, we've saved the best until last.

Should we ever want to truly make like Clark Kent's alter ego, it will be Yves 'Jetman' Rossy who gets a call from us. His set-up involves a high-powered jetpack with wings, meaning that he genuinely looks like Superman (minus the tights) soaring horizontally at incredible speeds through the air. It's so impressive that if you saw him flying overhead we reckon people around you might shout those iconic words 'It's a bird! It's a plane!' And now you'd know that the correct reply is: 'No . . . it's Jetman!'

He-Man – Superhuman Strength

Childhood just wouldn't have been the same without He-Man and good old BattleCat fighting evil Skeletor to protect Planet Eternia. Each and every glorious week, as we settled our little bottoms onto pillows and hunkered down to watch the latest episode, the Master of the Universe would hold his magic sword aloft and bellow 'by the Power of Grayskull!' – while mini-us did exactly the same thing playing along in our living rooms.

The key to He-Man, or Prince Adam's, success in warding off the evil Skeletor was his indestructible skin, super-speediness and, most famously of all, his superhuman strength. He could lift mountains or icebergs and even break photanium – the fictional strongest metal in the He-Man universe. He-Man's

---

* In George Orwell's world famous work of dystopian fiction *1984* there is a brilliant description of a journey in a solo-seater, personal helicopter along sky motorways' arteries over the suburbs of west London, which he imagined – in 1948 – would be in existence by 1984. This invention is well overdue!

super-strength came from the magical powers of Castle Grayskull, but how close have we humans come to attaining this seemingly impossible ability?

As a few hours spent watching any strongman competition will quickly confirm, many people spend their lives obsessed with trying to get there. For most, the obvious route is plenty of hard work pumping iron, along with, in many unfortunate cases, hefty doses of steroids. Not quite superhero-worthy . . . besides, the potential for serious injury and ill-health side effects is huge. So we're more interested in approaches that realise the ambition of making like He-Man without risking threat to life and limb.

One interesting phenomenon is that of average Joes and Janes reported to have suddenly displayed freakish bouts of seemingly superhuman strength in emergency situations. Like Oregon sisters Hanna and Haylee Smith who, in 2013, aged sixteen and fourteen respectively, lifted up a tractor to save their father who was trapped beneath it. Or twenty-two-year-old Lauren Kornacki who single-handedly lifted a BMW off her father to save his life back in 2012. Could understanding these kinds of freaky events be the key to making like He-Man?

The explanation for these displays of momentary superhuman strength, often referred to as hysterical strength, is that huge amounts of adrenaline and noradrenaline are released when people find themselves in life-threatening scenarios. The perception of danger also triggers a massive release of endorphins – a different set of powerful hormones that block pain messages passing from the body to the brain at the level of the brainstem. These also make us feel a bit euphoric, putting us in a state of mind that means we feel capable of anything, a bit like superheroes. Under normal circumstances, we are naturally prevented from exerting ourselves to the point where we start ripping muscles, or detaching the tendons that anchor muscles to the skeleton, by the pain that kicks in to warn us that we are on the verge of damaging ourselves. The flood of endorphins along with other hormones can release the natural brake on the forces we can apply through our muscles so that, rather than being limited to a fraction of the absolute maximum force that our muscles can produce to avoid causing damage, we can actually lift an estimated six to seven times our body weight.

There is superhuman strength in us *all*. Andy Bolton, watch out!*

However, impressive as they may be, these feats of hysterical strength only appear on the spur of the moment and in the context of an emergency, but for He-Man-style power status we need strength we can rely on 24/7. So, on our quest for the ultimate super-strength we turn our attention to mechanical exoskeletons.

In the classic film *Alien*, Sigourney Weaver climbed into her very own mechanical exoskeleton to do battle with the ugly, drooling, multi-mouthed extraterrestrial that stowed away aboard her spacecraft. Well, it turns out that this type of strength-enhancing exoskeleton now exists in real life and you can see them in action on Geoje island, just off the south coast of South Korea. There, in the Opk-Dong shipyard, there are workers who are using these mechanical exoskeletons on a daily basis, possibly at this very moment. The suits are made from aluminium, carbon and steel, which enables them to wield heavy lumps of metal with great ease. We thought driving a crane or a forklift truck for a living was impressive, but crouching down, reaching out with your metallic arms and picking up a blacksmith's anvil like some kind of superhero strongman must feel pretty incredible.

The only catch with these metal suits, made by Daewoo Shipbuilding and powered by hydraulic joints,† is that they enable any old weakling to carry just 30 kilograms without putting any strain on the person's body whatsoever. While this does satisfy the important need of reducing wear and tear on workers' bodies, it doesn't reach our He-Man super-strength dreams. As worthy a project as the exoskeleton might be, we've seen plenty of meatheads at the gym lift more than 30 kilograms. Even those who *aren't* on steroids can often bicep curl 30 kilograms without too much trouble.

---

* Andy Bolton is currently the world record holder for powerlifting, managing to hoist a whopping 550.5 kilograms above his head whilst in a squat. That's similar to picking up three Dr Jacks or four Llianas. It gives us sore thighs just thinking about it!

† With hydraulics, the trick is that by forcing liquid down a series of tubes you can create movement at a distance, like when you hit the brake pedal when driving and it transfers this motion to the mechanism that actually makes the wheels slow down.

It is a promising start, though, and Daewoo are promising to up the ante to 100 kilograms in the near future. With a few other contenders out there building similar and possibly stronger exoskeletons and then posting videos of them online, we'll certainly be keeping our eyes open for the latest developments. Because we'll only be truly satisfied that we've achieved He-Man levels of strength when we can lift a metric ton, and, by the power of Grayskull, mechanical exoskeletons are surely our best bet when it comes to reaching this superhero goal in the real world.

> **GEEK CORNER:** While we're waiting for the exoskeletons to sort their act out, you may find the battle between the Japanese Kuratas and America's Megabot fighting robots diverting. It reminds us of the classic film *Robot Jox*. Maybe someday mankind really will dispense with old-fashioned wars in which millions perish and instead just pit one country's human-driven, giant, armed, military robots against another.

Johnny Mnemonic – Super-Memory

Johnny Mnemonic may well have been one of those films that nearly passed you by completely unnoticed. If you did finally get around to watching it you'll have seen various gangs and corporations dead keen on chopping Keanu Reeves's head off to get their mitts on the precious data stored inside his brain. With no knowledge of what was actually stored on the 160GB chip* wired into his brain, Johnny (Keanu Reeves) was essentially just a walking, talking mule. But, rather than sneaking narcotics across borders in his personal passages, the contraband he was smuggling in his brain chip was information about the cure for a disease that had infected the whole planet.

CHIC FACT: In *Johnny Mnemonic* the fictitious disease infecting the planet was called 'neural attenuation syndrome', caused by over-reliance on smartphones, laptops and tablets. Just imagine! Over-reliance on smartphones? Where do those Hollywood script writers get these crazy ideas from?!

As a side effect of having a computer chip fitted beneath his skull Johnny had no access to his childhood memories, which is a pretty big price to pay, we're sure you'll agree. In real life we haven't got anywhere near the level of sophistication of Johnny Mnemonic's memory chip, but we're certainly making headway in helping to restore lost memories.

DARPA's real Restoring Active Memory project, launched in 2014, aims to insert microchip implants to boost people's memories and eventually help the two million Americans, including 270,000 wounded soldiers, suffering from traumatic brain injuries every year. The technology is currently being tested on a brave batch of volunteers who already need brain operations for other neurological conditions. They agreed to have the experimental implants fitted in brain areas involved in creating and recalling memories relating to facts, figures and spatial navigation. This enabled scientists to do vital research on the neural code that brains use to encode memories and the appropriate timing of electrical pulses required to actually improve memory consolidation and retrieval. It's still early days, but preliminary findings are promising, suggesting you can improve memory recall by targeted electrical stimulation of the brain. Great progress is being made in this rapidly growing area of

---

* Incidentally, the 'wetware' chip embedded in his head was disguised as a standard-issue device for fixing dyslexia – a concept that we think is ace!

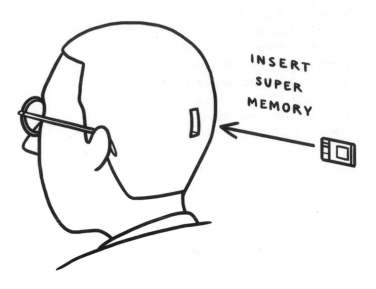

INSERT
SUPER
MEMORY

neurotechnology and we'll be keeping our beady eyes on these promising projects as they continue.

Meanwhile, Canadian researchers in 2014 discovered a molecule that may well be the key to unlocking 'super-memory'. Every day thousands of things happen and yet we will only ever remember a few, if any, of them. Without the selective filing away of pertinent memories our brain cabinets would simply get full up and potentially malfunction. The process of selecting which memories to retain and lock away and which ones to discard is a complex one, reliant on many factors, but one of the contributors is a protein known as FXR1P. FXR1P actually limits memory formation, by suppressing the production of other vital molecules so that events not deemed worthy of our later recollection are simply forgotten. The research team removed this protein from certain parts of mice brains and found that the memory-creating molecules were produced in much higher quantities. The result was much better memory and recall. Bingo! Of course, further research is needed to see whether adapting this to human brains could result in total recall or some kind of super-memory, but the study is promising, especially for research into Alzheimer's and autism as well as for us superhero wannabes.

GEEK CORNER: A new Spider-Man-style form of brain electrode has been invented that is made out of a biocompatible polymer and injected deep into the brain via a small tube. This then unravels like a spider's web inside the brain, forming a scaffold that neurons can actually grow through and interface with.

In early 2016, another amazing breakthrough occurred, allowing people to gain direct access to other people's knowledge, memories and skills, learning from their example with no extra effort required on their part. Brainwaves of commercial and military pilots were recorded using EEG (electrodes that sit of the surface of the scalp as opposed to inside the brain) and then transmitted into the brains of novice pilots using transcranial direct current stimulation (tDCS). This improved the newbie pilots' ability by 33 per cent, meaning they flew simulators more like the pros without requiring nearly as much training. We'd love to tap into these *Matrix*-style learning gains, but without the need for a cumbersome tDCS machine to stimulate our brains.

CHIC FACT: Some people have natural super-memories, and can remember practically every day of their lives. Around fifty people are known to have a highly superior autobiographical memory, including ten-year-old Jake Hausler. Jake and several others with super-memories have had their brains scanned by MRI and were found to have more active pathways between the back and the front parts of the brain. The only downside to Jake's 'superpower' is that he vividly remembers all the bad memories as well as the good.

Progress is being made towards achieving the super-memory dream and, as ever, it's science that is making the critical difference. Microchips are certainly the current front runners but, should we ever find ourselves inserting them into our brains, we'd want to see some immediate, tangible and astonishing benefits. We're thinking of a microchip (delivered to your door by Amazon drone – in thirty minutes or your money back) that we can plug directly into a socket embedded in our head to give us instant access to mankind's knowledge. That way we'd never find ourselves hindered by a weak or non-existent Wi-Fi signal when we need to access the seemingly infinite wisdom of the internet. Omniscience on a chip. Sweet deal.

Aquaman – Underwater Breathing

Picture this. You're 30 metres under the sea, enjoying the scuba dive of your life and surrounded by the most beautiful marine creatures and multi-coloured corals. You're feeling as serene as can be, but then as you glance down to the gauge for your air tank you suddenly realise . . . shock, horror! Panic stations! You're running on empty.

It's every scuba diver's worst nightmare and just the thought of it is enough to bring us out in cold sweats. So all hail the advent of a wondrous invention that might well soon make this horrible possibility a thing of a past.

> **CHIC FACT:** Two hundred people die every year while scuba diving and one survey found that 41 per cent of dive fatalities are caused by running out of air. That's eighty-two lives the Aquaman crystal could save every single year, so this might just be the best candidate for helping us realise this particular superhero dream.

It's all to do with a revolutionary new lab-made crystal, dubbed the 'Aquaman crystal', after many people's favourite underwater hero (although personally we would have plumped for the Little Mermaid crystal). Originally developed by Danish scientists to help people with respiratory problems, its secret is that it contains cobalt. As you may remember from your school biology lessons, certain metal atoms have an incredible capacity to absorb and transport oxygen, then release it when and where it's needed. In us humans it's the iron in the haemoglobin of our blood that helps to carry oxygen around our bodies, whereas in crabs and spiders it's copper that does this vital job.

The clever Danish scientists who invented the Aquaman crystal may well have taken a sneaky peek at the periodic table and noticed that, as cobalt sits between copper and iron, it might just have similar oxygen-carrying capacities. They created cobalt-based crystals and it turned out their conjecture was correct! The crystals *were* able to absorb oxygen from the water or air around them, store it and release it when needed. What's even better is that only a few grains are needed for one single breath.

CHIC FACT: The Aquaman crystals will also be developed for medical uses, helping lung disease patients who would normally have to carry big, cumbersome tanks around with them.

But here's the best part for the scuba-loving bods of the world: once all the oxygen is used up the crystal can simply replenish its store from the surrounding water and repeat this process in a continuous loop. It's a bit like a sponge, except with oxygen instead of water being sucked up and squeezed out over and over again. And how exactly do we 'squeeze out' or release the oxygen when we need it? Right now it takes heat, or low oxygen pressure, like that found in a vacuum. But researchers are hoping soon to develop a way to release it with light, bringing us one step closer to realising our dreams of becoming real-life water babies.

CHIC FACT: The Guinness World Record for the longest time someone has held their breath underwater is twenty-two minutes, currently held by Stig Severinsen.

The only tricky bit is being able to replace the oxygen as fast as you consume it, but the teams are working on making this improvements. Once they've achieved this, scuba divers across the world can quite literally breathe easy, knowing that soon they will have a never-ending, self-replenishing oxygen store in the form of sacs of tiny crystals; much lighter and easier to carry than conventional scuba tanks.

We do feel bad for the Little Mermaid, though; she had to give up speaking to breathe on land, whereas to breathe underwater we'll only have to sit back and let the tiny metallic element called cobalt work its magic.

Iceman – The Man Who Feels No Cold

Do you remember *Marvel*'s Iceman? We reckon he's a bit of an understated legend. He always kept cool in tricky situations and from his outstretched palm magical silver slush could be jettisoned, which miraculously solidified the very moment it made contact with air, enabling him to skid along a never-ending roller coaster of his own creation.

Clearly, a prerequisite for this role is an amazing tolerance for the cold. And while real-life 'iceman' Wim Hof has never quite mastered the sleet-flinging technique, he has nailed the freezing cold tolerance part. He can swim under the Arctic ice on a single breath of air for over five minutes, or walk up to the very top of Mount Kilimanjaro in a pair of shorts and little else. He developed a number of techniques for switching off the perception of pain that we normally feel when our skin is exposed to very cold temperatures. In the process, he accidentally uncovered some strategies for exerting control over his autonomic nervous system and, excitingly for anyone suffering with an autoimmune disease, dampening his immune system, too.

In a quest to share his incredible powers with others, Wim Hof got scientists involved to try to figure out what actually happens inside his body to enable him to withstand such freezing temperatures.

What the scientific studies at Radboud University in Nijmegen in the Netherlands showed was truly astonishing. Wim's hyperventilating breathing tactics enable him to blow off so much carbon dioxide* that his blood actually becomes less acidic. This is because carbon dioxide dissolves in the blood to form carbonic acid, which keeps the blood's pH relatively low. By getting rid of more of the carbon dioxide through rapid breathing, the carbonic acid concentration of the blood is reduced, which pushes the pH up to levels that just happen to inactivate pain receptors in his skin.

GEEK CORNER: At normal pH levels a critical protein in the pain receptor, made up of three component parts, fits together in a very specific 3D configuration. At the high pH levels found in the blood of Wim, and anyone who regularly practises his techniques for that matter, these three parts can't fit together properly, so the pain receptors can no longer send electrical messages to the brain that would usually be interpreted as feeling pain. When plunging themselves into icy water they still get the sensation of extreme cold – which Wim uses to get himself into a meditative trance – but without any of the discomfort usually associated with such temperatures.

Better still, they discovered that when inactivated endotoxins were injected into Wim's bloodstream, which would make most of us break out into a fever within twenty minutes, his immune system simply didn't respond in this way. He just sat there chilling as if nothing had happened. Why? Because the combination of breathing techniques and cold immersion somehow changed how his immune system responds to the invasion of foreign bodies. Turns out, the interleukin (IL) profiles in his blood are different from the rest of ours, though we can change ours too, if we so wish, through training – officially called the 'Wim Hof method'. Specifically, levels of

---

* We produce carbon dioxide in our cells every moment of every day as a waste product of the mechanism by which our bodies release energy from glucose – this needs to be constantly removed so as not to build up to toxic levels but is removed more quickly if we breathe in and out quickly and deeply.

the pro-inflammatory substances IL-6 and IL-8 are reduced, while the level of anti-inflammatory IL-10 is increased. The combined effect of this is a less reactive immune system. Which at first glance, we agree, seems like a terrible idea!

Why in the world would anyone want to make their immune system *less* effective? Well, several diseases of the modern age – such as rheumatoid arthritis and multiple sclerosis – are autoimmune diseases. In other words they result from an *overactive* immune system that starts turning on its own healthy cells. In rheumatoid arthritis the target of this assault is the soft tissues in the joints, which become painfully inflamed and eventually, after many years, physically deformed. In multiple sclerosis it is the central nervous system that comes under attack, damaging the white matter* that enables different areas of the brain to communicate with each other and the body.

Now that the Wim Hof Method has been scientifically accredited with having a genuine impact on immune system function, and proven to work in anyone who follows Wim's technique, there is a huge interest in its therapeutic potential.

Many a crank has tried to convince the world that they have genuine superhero powers, but they always seem to be determined to keep all the adulation for themselves. *Me, me, me, me! I am the master!* The amazing thing about the real-life Iceman isn't just his incredible combined superpowers of withstanding intense cold and immune system suppression through scientifically verified practices of mind over matter, but the fact that he felt compelled to share his technique with others. By using science to unlock and then give all his secrets away, for free, Wim Hof's heroics win him the coveted prize that is our just-invented Geek Chic Superhero Cup. After all, pure altruism has to be the ultimate superhero trait. Mr Iceman – we salute you.

Much as superhero abilities are great, we reckon there are a couple of other places we can look to gain the powers for world domination . . . err, we mean saving the world in a superhero manner! We're one of the good guys (and gals), honest.

---

* White matter is the parts of the brain made up of bundles of axons and connecting the grey matter.

The Invisible Man and Invisibility Cloaks

Making an invisibility cloak is really easy. All you need is a normal, common-or-garden cloak and the hair of a Demiguise – you know, the magical creature that possesses the power of invisibility. Simply weave the hair into the aforementioned cloak *et voilà* – don this incredible garment and you will miraculously disappear from sight. But be warned: over time the hair will eventually start to become opaque at which point the cloak starts to lose its powers of invisibility. Good thing that Xenophilius Lovegood warned Harry, Ron and Hermione of this before they—

Hang on, hang on – wrong book! We're not in the land of J. K. Rowling, are we?

Good news, however, because back here in the real world of hard science (which admittedly can sometimes seem more magical and astounding than any work of fiction) we may be one step closer to creating real-life invisibility cloaks. Turns out that it may all be down to Mother Nature's very own Demiguise, a 'magical creature that possesses the power of invisibility' that has been swimming in our seas for millions of years. The name of this miraculous being? The humble squid.

Squid are one of only a handful of animals that can change the reflectance properties of their skin in order to blend in with their backgrounds. Creatures generally evolve this fantastical ability in the hope of avoiding being snagged by passing predators or, occasionally, to lure unsuspecting prey close enough to pounce on. Not only does this result in some rather beautiful videos of squid changing colour as they swim across tropical corals but it has also driven many scientists to seek inspiration from biology to aid their quest to create the elusive invisibility cloak.

In 2015, a research team from the University of California at Irvine identified the light-reflecting protein in squid skin responsible for their marvellous disappearing acts. Aptly named 'reflectin', it is contained within special cells called iridicytes, the light-reflecting properties of which squid can alter, as easily as we can flutter our eyelids, to rapidly change colour and match their environment.

The team decided to create bacteria that could manufacture reflectin in large quantities in a lab. They extracted the synthetic reflectin and combined it with an uber thin and special material called

graphene (see p.74) to layer it up onto a sticky-tape-type substance and create rolls of invisibility stickers, each around 100,000 times thinner than a human hair, which people like us could potentially stick all over our bodies or stuff to make them disappear.

CHIC FACT: The first real 'invisibility cloak' was unveiled in 2006, using specially designed meta-materials to bend light around objects (a little like water flowing around a rock), to prevent it from reflecting the electromagnetic radiation back. The problem? It only rendered objects invisible to microwaves. Useful in the world of communications, not so great for those of us wanting to make like the Invisible Man.

Sounds magic. However, before you run off to plot an orgy of invisible mischief, it's not quite as simple as that. While squid use a specially evolved biochemical cascade to trigger their reflectin colour changes, in the lab they had to use acetic acid vapour (essentially a strong vinegar), and we're pretty sure the Invisible Man didn't have to douse himself in stinky vinegar every time he wanted to disappear. However, the teams are working on ways to trigger the changes in the reflectin molecules using more convenient manipulations, like stretching or compressing the tape instead, but at the moment the research is still in its infancy.

Another minor problem is that, at the moment, these invisibility stickers only work in darkness, with the first planned use for rolls of these squid-inspired stickers being to make soldiers invisible to the enemy's night vision. When applied liberally to the surface of their army fatigues, the first-generation stickers should be able to disperse the infrared lights used to detect them, in a similar way to their surroundings, thus helping them to evade detection. Happy days for them. Not so good for our daytime superhero needs.

CHIC FACT: The army once made an entire tank disappear, fashioning an 'invisibility cloak' using tiny little cameras and mirrors.

There's still a fair old way to go before we achieve our Harry Potter dreams, but we remain ever hopeful for a squid-inspired future. But just remember, when you do finally get your invisibility cloak, make a mental note of where it goes when putting it away, otherwise there's a good chance you'll never find it again (ba dum dum!).

Having an invisibility cloak would certain help us travel about incognito, but what if we wanted to travel far and wide without any of the hassle of moving our lazy asses at all. Could we ever forget the passports and the long queues for check in and actually teleport ourselves somewhere exotic?

The Mysteries of Teleportation

The first time we came across teleportation was in *Rentaghost*, the eighties children's series set in west London just a few miles from where we both grew up. One of the characters – Nadia Popov – would magically disappear whenever she sneezed, reappearing at random somewhere else in the house. Frustrating for her, hilarious for everyone else. Fast-forward just over a decade and Nintendo released the classic punch-up videogame *Street Fighter II* where, alongside Ryu,

Ken, Chun Lee, Guile, E-Honda and others, the rubber-limbed, Indian yogi Dhalsim could unleash a number of extraordinary powers. One of these was his ability to jump up into the air, assume the classic cross-legged meditation pose and then promptly disappear in a flash of light only to reappear immediately behind the opponent to give him a swift punch in the back of the head.

Of course, for others their dreams of instant transportation will have come from decades of watching Captain Kirk muttering 'Beam me up, Scotty!' into his flip lid communicator. But how far have we come in realising the vision of *human* teleportation? Absolutely nowhere. Not a sausage. If you're interested in transporting physical matter from A to B, prepare to be disappointed.

Quantum teleportation, on the other hand, is a very different matter. Science has been successful in transporting the 'quantum state' of teeny tiny things like light beams. Which is pretty handy, it turns out, if you have a genuine need for both speed and privacy.

Professor Nicolas Gisin leads one team looking into this area and, in 2014, his laboratory smashed the distance record for quantum teleportation. Professor Gisin kindly recorded a *Geek Chic Weird Science* podcast special with us soon after and told us how his team had managed to teleport the quantum state of a photon, or particle of light, across very large distances. The key to this is to first separate two quantum-entangled photons.* Quantum entanglement is admittedly an incredibly difficult concept to grasp, but the way we like to think of it is by imagining the pairing of yin and yang. If one photon is in 'yin' state, then its quantum-entangled partner photon *has to be* in 'yang' state. If the original photon suddenly changes to from 'yin' to 'yang' state, then the partner photon will spontaneously switch from 'yang' to 'yin' state, *no matter how far apart they are separated*! Don't ask why. Don't ask how. The answers to those questions are at the very limit of scientific understanding. For the time being, if you can just accept that this relationship between quantum entangled photons always runs like this, then you might just be able to wrap your head around how useful this phenomenon could be once fully exploited.

Once photons have been separated, changes in one photon are

---

* Quantum entanglement is a rather complex quantum property that means particles can communicate across infinite distances.

instantly registered in its quantum-entangled partner proton no matter whether it is one centimetre or one light year away. While this may not be an *Enterprise*-style transporter, it might be just the ticket for sending instantaneous and unsnoopable messages across potentially unlimited distances.

So far, this feat has been achieved using photons over a separation of just over a hundred kilometres, but in theory it could allow instantaneous communication across much larger distances – such as between Earth and Mars, for example, should NASA get their way and set up a human colony on the red planet (see pp.67–71). If that did become a reality, one massive ball-ache with our current set-up would be the need to wait between four and twenty-four minutes for any message sent from Earth to reach Mars,* or vice versa, which hardly lends itself to a snappy dialogue for the WhatsApp generation. But if a few quintillions of quantum-entangled photons were separated from each other, with half left on Earth while their partners were transported to Mars, then you might just be able to use them to detect changes in the quantum state at the far end of this futuristic cup-and-string quantum telephone, as they would instantly be registered at the other end, some 250 million miles away.

> **GEEK CORNER:** Explaining how a change of state in one photon is instantaneously registered in its partner across vast distances requires mind-bending concepts like wormholes to connect the two via another space-time dimension, so we're not even going to go there. Apart from saying that YouTube sensation The Physics Girl has a video with a lovely demonstration of a 'crazy pool vortex', which you can recreate yourself in any swimming pool with a dinner plate and some dye. It illustrates just one of the many invisible forces operating right under our noses every moment of every day. Just imagine what goes on when it's at the unfathomably tiny scale of all things quantum!

---

* The difference in the waiting time depends on the constantly changing distance between Earth and Mars as they independently orbit the Sun.

In reality, the first application of quantum teleportation is likely to be much closer to home. A quantum internet could pass messages across the ocean with zero time lag, no possibility of eavesdropping and no chance a fishing trawler might cause trouble by snagging underwater cables. Why? Once the cables have been laid across the ocean to move one half of each quantum-entangled photon pair to the other side – the cables can then be promptly removed. With no actual physical wires connecting one photon to the other, there's no way of knowing what's being communicated. Sneaky.

While we may be a tad disappointed that our dreams of making like *Street Fighter*'s Dhalsim or demanding 'Beam me up, Scotty!' are still well beyond our reach, quantum leaping data is still a pretty mind-blowing concept.

As to whether a living being could ever be teleported, Professor Gisin told us on our podcast that we can never rule anything out, saying:

> We know how to teleport structures of atoms and photons and this same technology that we are mastering now will not apply to large objects, let alone to a virus, much less to a human being. So, it's not just a matter of developing new technology, we need new ideas and it's impossible to predict when new ideas may emerge.

Call us optimists but we'll take that as a maybe. Clicking your fingers to get to the Bahamas may only be one great idea away from becoming a reality.

A Trekky-style transporter is unlikely to exist in our lifetimes, if ever, but it's clear that popular culture has an influence on the direction of scientific experimentation (and vice versa!). And Star Trek-*inspired research isn't limited to teleportation – the famous Mr Spock mind-meld may soon be within our grasp.*

Mr Spock's Mind-Meld?

Mr Spock, son of Sarek, was the Starship *Enterprise*'s science officer in the original *Star Trek* series. His pointy ears weren't the only thing that made him special. His Vulcan blood enabled some mighty impressive acts of telepathy – the ability his fictitious species is

probably best known for. Vulcans perform a type of hands-on telepathy that allows them to merge their minds with another being – a spectacular form of privacy invasion more commonly known as the mind-meld. Edward Snowden wouldn't have been a fan.

Once successfully telepathically linked to another person, memories, emotions and experiences can be shared between melded minds. Over the course of many series of *Star Trek*, various Vulcans performed mind-melds with many different species including humans and even humpback whales. The question that immediately rears its curious head at this point is: can science make it so?

> **CHIC FACT:** The late, great Leonard Nimoy, who famously played Mr Spock, was not the first choice to play the Vulcan superstar. DeForest Kelley was offered the role first but turned it down to play Dr Leonard 'Bones' McCoy instead.

Using the latest technology, the answer is: sort of. Certainly it's not yet possible for one person to lay their hands on the head of another and, with a little bit of transcendental meditation, suddenly find themselves knowing what the other person is thinking. But stick someone in an MRI brain scanner, give them a series of pictures to look at and it's a different story. Using a sophisticated algorithm, it's possible to recreate a crude but recognisable reproduction of the image the person in the scanner is looking at, based purely on their brain data. That's pretty cool but, as we're sure you'll agree, it's a long way from actually reading someone's mind.

What about reading the thoughts of another person over a great distance – on a different continent, for example? Even the mighty

Mr Spock couldn't perform a full mind–meld further away than arm's reach, so surely it's beyond our own feeble powers? Surprisingly, given the right tools, it is already entirely possible for one person to think directly into another person's brain at pretty much any distance, as long as both have an internet connection and a bit of time on their hands.

Here's how it works: the sender sitting in a lab is fitted with an EEG head cap, which monitors changes in the electrical field produced by their brains, via the numerous electrodes held against their scalp. A binary code equivalent to the letters of the alphabet is then established, e.g. an 'h' might be encoded as '0-0-1-1-1' and an 'a' could be '0-0-0-0-0'. The sender is told to imagine performing a movement with their hands to signify a '1' or imagine moving their feet to signify a '0' – the EEG picks up which one was being thought of according to the different brainwaves generated by each. By imagining sequences of movements in their hands or feet the sender is eventually able to spell out words like 'hello' and 'ciao'.

These message were then transmitted (via email!) in binary code to a lab in a different continent, where a second punter (the receiver) is blindfolded and rigged up with a transcranial magnetic stimulation (TMS) device resting against their head. Applying current through this non–invasive TMS device produces a localised magnetic field that triggers brain cells at the back of the brain where visual perception takes place. So, this TMS stimulation produces 'phosphenes' in the person's mind's eye – flashes of spooky rings of light which seem to appear and disappear out of nowhere in their peripheral vision.

CHIC FACT: Mr Spock was originally supposed to be half-Martian, with red skin, but as the series was originally in black and white the execs realised his face would look too dark on screen and plumped for a lovely yellow hue instead.

The receiver knows that a flash of light means '1' and that no flash means '0', so they can eventually receive the binary message and decode it to turn it back into text. Successful tech–enabled telepathy – even if it does take seventy minutes to relay one simple word!

The scientists responsible for this pivotal research made headlines all over the world with the first ever intercontinental, direct 'brain-to-brain' communication.

Playing devil's advocate for a moment, we could argue that, seeing as in this particular study the message was sent from India to France via email, surely a more straightforward method would be, you know, just to type an email. Still, being able to connect human brains to each other directly in order to share thoughts without use of spoken or written words is an exciting development all the same, even if it is only a proof of principle. It is a development that could potentially be a huge help to patients unable to communicate with the outside world, but whose brains otherwise function just fine, such as those with locked-in syndrome. We have a sneaking suspicion that Mr Spock would be proud.

While brain-to-brain communication may one day help patients with locked-in syndrome, elsewhere in the medical profession huge leaps are being made to revolutionise the treatment of paralysed patients. It takes an altruistic person to choose healing others as their superpower, but helping others is what some incredible people do each and every day.

Apollo – Super-Healing

The physician must be able to tell the antecedents,
know the present, and foretell the future – must mediate
these things, and have two special objects in view with regard
to disease, namely, to do good or to do no harm.
Hippocrates (aka the Father of Medicine), fourth century BC

Sure, Superman may have looked hot in his little blue onesie, but Hippocrates and his followers have saved many more lives. And we reckon the ancient Greeks had it right when they worshipped the god of healing, Apollo, because in real life it's the doctors who are the heroes of the day. They work incredibly long hours, dedicate their lives to helping others and regularly pull off seemingly miraculous feats – and none more impressive than in the incredible case of Polish fireman Darek Fidyka.

Mr Fidyka was left paralysed from the chest down after a knife attack in 2010, during which he was stabbed in the back several times, almost completely severing his spinal cord. As if not being able to move his legs was not bad enough, he became unable to control his bladder and bowel movements and was robbed of his capacity to engage in sexual activities. At the time of Mr Fidyka's accident the conventional wisdom was that it was impossible to repair a completely severed spinal cord. Yet today he can cycle around on a tricycle, walk (with support), has conquered his incontinence *and* regained his sexual mojo. How? The answer of course, is our heroic super-doctors.

> **CHIC FACT:** The £250,000 per patient treatment is funded by the UK charity the Nicholls Spinal Injury Foundation, as long as candidate patients are prepared to spend three years in Poland for treatment.

A skilled surgeon named Pawel Tabakow snipped out part of one of Mr Fidyka's olfactory bulbs.* Precious olfactory ensheathing cells (OECs) contained within the bulbs were then extracted, multiplied in a dish and then later reintroduced to his body just above and below the eight-millimetre gap in his spinal cord, to encourage regrowth of the severed fibres.

---

* We all have two olfactory bulbs, one above each nostril, through which odour-related information enters our brains via neurons that pass through special perforations in the section of the skull that separates the nasal cavity from the brain.

GEEK CORNER: Around 500,000 OECs were cultured from Mr Fidyka's olfactory bulb and injected into his spinal cord via one hundred micro-injections above and below the site of injury. This all occurred just two weeks after the original operation to remove one of his two olfactory bulbs. Four thin strips of nerve tissue were also taken from his ankle and laid across the gap in his spinal cord to give the OECs a scaffold to grow across.

This procedure followed decades of ambitious neuroscientific experimentation: various animal models had first been used to test this novel approach to treatment for severed spinal cords, then a few brave volunteers took part in human trials to make sure that the procedure was safe enough to fulfil the Hippocratic edict 'Do No Harm'.

However, this immense team effort eventually resulted in monumental success with Mr Fidyka becoming the first person in the world ever to recover from complete spinal cord separation*. There's hope for others, too. As of March 2016, anyone aged between sixteen and sixty-five with permanent paralysis due to a spinal cord injury can apply to the Wroclaw Walk Again Project in the hope of becoming the next test patient for this fantastically hopeful new therapy.

We can't help but think, as we muse on the superpower of healing, that Christopher Reeve's own doomed struggle with paralysis makes Mr Fidyka's ultimate triumph bitter-sweet. Christopher Reeve, who played Superman in the original films, tragically became paralysed from the neck down after being thrown from a horse, never to walk again. For years he ploughed huge sums of money into experimental treatments for paralysis in the hope that it might one day help him, and others who found themselves in a similar predicament, to walk again. Sadly, he died in 2004 from a cardiac arrest, over a decade before he would have seen the scientific breakthrough he had been

---

* This surgery only serves to plant the seeds of repair and a huge amount of additional effort is needed by the patient, before and after the procedure, which comprises up to five hours of physiotherapy daily to promote regrowth of nervous tissue across the gap.

striving for all that time finally bearing fruit. His legacy lives on, however, not only as Superman, but in getting the ball rolling for the miraculous treatments that are the embodiment of the ultimate altruistic superpower: healing other people.

> **GEEK CORNER:** Olfactory ensheathing cells enable continuous replacement of the olfactory receptors on the inner surface of your nasal passages. When specific gaseous chemicals inhaled through the nose bind with these receptors they send a torrent of electrical messages to the brain. These neurons pass through special sieve-like perforations in the base of the skull, just above the nasal passages, called the 'cribiform plate', enabling them to make synaptic connections with other neurons in the brain's olfactory bulbs responsible for actually generating the smell of a scent.

By now, hopefully you've realised that, thanks to a whole load of innovative technology, a few brilliant scientific minds and a little sprinkling of good genetics, some superhero abilities are well within our grasp. Happy cape-wearing days ahead for us, then. But – and here's the caveat – it's a lot of effort, isn't it? Being able to fly, cope with freezing temperatures, implant thoughts into others' minds, lift huge weights, breathe underwater, help paralysed people to walk again and so on is pretty exciting, but who's really got all that time, money and energy required to invest in it? What if it was all available to you at the flick of a switch?

It's a Bird! It's a Plane! No, it's Virtual Reality!

Wouldn't it be nice if there was a way to have *all* the superhero powers you could ever dream of, but without any of the blood, sweat and tears it takes to get them? Well, guess what, it's your lucky day, because you can. Sort of. In a virtual world.

Virtual reality (VR) has been around for some time now, but as the years go by technology gets more powerful and immersive virtual worlds become ever more real. At the moment the front-runners are HTC Vive, Oculus Rift and Samsung's headsets, which you place

over your eyes and, in combination with a suitably powerful gaming PC, they can grant you all sorts of virtual superpowers.

Some of them let you fly around like a superhero – there's one where users point their arm forward (Superman-style) and whizz around cities, saving a few lives here and there. Or if being Batman has always been your childhood dream then Oculus Rift might just be your new best friend. Their Batman-style virtual world allows you to follow in the footsteps of Bruce Wayne, exploring the bat cave, Gotham City and driving the Batmobile itself to carry out all sorts of ridiculous crime-fighting missions.

Another VR headset allows you to become jet-packing secret agent, flying around a 3D cartoon world. And in case that isn't quite sexy enough, your jet pack can also use its built-in missiles and guns to destroy UFOs. Nice. Alternatively, if you want to feel like an actual bird there is a frankly ridiculous looking contraption named Birdly that allows you to lie face downwards, flap your arms like bird with some plastic wings, with a fan blowing wind and various landscape-related smells into your face. Apparently it feels amazing, but you do look like a bit of a berk.*

The possibilities are quite literally endless and virtual worlds are only going to get more realistic and commonplace as time goes by.

---

* 'Berk' is a Cockney rhyming slang term, but it's far too rude to translate here . . .

However, quick confession here: neither of us are really big gamers (sacrilege for self-confessed geeks, we know). But – and here's the part that really floats our geeky boats – some studies have shown that what you experience in a virtual world can feel so real that it can have serious psychological and behavioural effects back in reality.

For example, if a person walks over a log perilously dangling over a big drop, the virtual walker may know rationally that they aren't going to get hurt, but the illusion is so strong that their stress levels will rise sharply all the same. On the flip side, people with phobias about flying who experience going on a virtual plane could overcome their fears in real life.

One experiment even decided to test whether giving someone virtual superhero powers would make them start acting more hero-ically in real life. Participants got to either virtually fly around like Superman or virtually fly in a helicopter (talk about getting the short straw). Both groups of 'flyers' were given one of two tasks, either touring a virtual city (snooze) or helping to find a missing child des-perately in need of help ('here he comes to save the daaaaay, Mighty virtual-Mouse is on his waaay' – that's more like it!). Now here's where it gets clever, because the real test started once all the VR fun was over. Once the participant finished their flight and removed their VR headset the experimenter would 'accidentally-on-purpose' spill a load of pens all over the floor. So, who was most helpful? Was it the virtual-helicopter passengers? No, of course not! Those luxury-loving fools had no desire to be altruistic whatsoever. The superhero flying participants were by far the more helpful, whether they'd toured the city or found missing kids. Our experimenters' conclusion? Being able to virtually experience the 'superpower' of flight makes us more likely to be helpful back in the real world.

We can't wait to see future experiments showing how other virtual superpowers could affect our behaviour in real life. Would acting like virtual Bananaman finally make us like bananas?

Final Thoughts

We may have *slightly* exaggerated in our introduction to this chapter when we promised you the Moon on a superpower plate and sug-gested science had cracked *all* the most coveted superhero secrets.

However, progress is steaming ahead and various scientific communities are making incredible headway in granting us our wishes and developing technologies to enable the abilities we grew up thinking would remain impossible forever, to become well within our grasp.

Hmmm ... maybe it's the scientists behind all this who are the real superheroes after all.

Acknowledgements

First and foremost thanks to you, our lovely reader, for buying and reading this book. You rock. Also to all our loyal listeners who've been with us throughout the podcast adventure, Geek Chic's Weird Science, that this whole book is based on – without your continued support this book quite literally would never have been written.

A huge thank you also to Richard Boffin, who for nearly three years now has skillfully managed the sound production of our Geek Chic Weird Science podcast without payment, complaint, nor skimping on his high technical standards.

We are also greatly indebted to Rhiannon Smith for doing an absolutely fantastic job as our editor, and to Adam Strange and Adrian Sington for negotiating the book deal in the first place, not to mention Radio X for letting us use their brilliant facilities for our podcast recordings.

Finally to our Matt Blease for his witty illustrations throughout the book, to the Noel Fielding for the most excellent painting that graces our cover, and to Dave Brown for making it all look super hot with his design skills – we thank you all.

Lliana's personal acknowledgements

First and foremost I'd like to thank my writing and podcast-making partner in crime, Jack, for his dedication and continued passion (and for waiting for me patiently every single time I was late for every single one of our hundreds of meet ups!). From the moment he first joined me as a guest on my radio show for our first trial run of a mini 'geek chic weird science' feature, to the very last word written in this book, he's been a wealth of science knowledge and above all, a true friend.

I'd also like to thank my sister Tania, for being the brilliant scientist she is and encouraging me to discover more, to my dad for passing his strange logical brain on to me and for inspiring my curiosity and thirst for science knowledge in the first place, and to my wonderful mum for making me the person I am today.

Huge love and thanks to my best friend Phoebe, who painstakingly read every single word of this book and gave us her considered feedback, even as my beautiful god-children Lula and Indy danced around us.

A special thank you to my literary agent Adrian for taking a punt on me, to the amazing Dawn for encouraging me to write in the first place, to my science teachers at Bristol University and Goldolphin school, and for Chris Baughen and the Radio X crew who (for some reason) continue to have be rabbiting away on air week after week. And the biggest debt of gratitude to my beautiful Josie Jo and all the incredible Help Refugees uber ladies for allowing me to be totally absent for large chunks of time as I battled to finish this book. You girls saved me.

Last, but certainly not least, to my partner in life and love. Not only for providing the brilliant cover to this book but for making me believe I can do anything, and then holding my hand every step of the way, from radio shows, to podcasts, to books, to starting a charity, to meltdowns, to moments of joy, despair or introspection. But most of all for always holding up a mirror to me and making me realise when I'm being a total brat. I love you.

Jack's personal acknowledgements

Firstly I'd like acknowledge Lliana for suggesting this collaboration in the first place. More importantly I'd like to thank her for being one of the few people in life who ALWAYS delivers on her promises. Being prepared to work hard, consistently, over many months, for zero remuneration is extremely rare. Studies have shown that work done in the context of a social contract is invariably much more rewarding than anything done under a financial contract. As our podcast has been a labour of love from beginning to end, I think this proves the veracity of this important insight. Thanks Lliana for your intelligence, wit and tenacity.

We've already dedicated this book to our parents but I feel the need to acknowledge them too, thanks to both of you for the time, effort and support you put in to getting me set up for life in the 21st century. Speaking of which, I'd like to thank all my teachers. Those at Latymer Upper School for helping me prune my brain's less important synapses and beef up all the really important ones by nourishing my love of knowledge, novel experiences and physical activity. Those at The University of Nottingham for providing me with such a solid foundation in neuroscience and for having the bravery to give some of my more risky and outlandish essays such extraordinarily high marks that I managed to scrape a 1st class degree. Those at University College London for arming me with the ability to explore the human brain using fMRI; and perhaps more importantly for developing in me the appropriate level of skepticism to be able to distinguish wheat from chaff when evaluating scientific papers. And finally my colleagues at The Max Planck Institute for Biological Cybernetics for the peace, isolation and exceptional facilities that enabled me to perform research of sufficiently high quality that it was deemed worthy of publication in a top science journal (at long last!). This was instrumental in helping me feel ready to turn my back on my fledgling research career forever in order to commit myself fully to communicating the breakthroughs in science, rather than trying to make them myself.

I would like to flag five friends in particular whose perpetual encouragement, over a period of decades, has kept me sane whilst boldly veering away from the straight and narrow in pursuit of my relatively unorthodox calling. Thanks a million to Tom Schroeder, Scott Fairbairn, George Wolstencroft, Angie Farrag and Dan Markham for the consistent provision of sage advice and unflinching honesty over so many decades. I'm also much obliged to Adrian Webster for agreeing to co-author my first book *Sort Your Brain Out* and for teaching me that: as hard as the process of merging two voices into one is to endure, it invariably leads to a better end product (so long as you manage not to murder each other in the process).

Lyric credits

'The Bad Touch': Bloodhound Gang, Universal Songs of Polygram International, Inc.; Hey Rudy Music Publishing, 1999. By James Franks.

'Chim Chim Cher-ee': from *Mary Poppins*, Wonderland Music Company, 1964. By Richard Sherman and Robert Sherman.

'Food Glorious Food': from *Oliver!*, Musicscope, 1960. By Lionel Bart.

'The Future is Now': The Offspring, Kobalt Music Publishing, 2012. By Bryan Holland.

'Heroes': David Bowie, Peermusic Publishing; Sony/ATV Music Publishing LLC, 1977. By David Jones and Brian Eno.

'It's the End of the World as We Know It': R.E.M., Warner/Chappell Music, Inc., 1987. By William Berry, Peter Buck, Michael Mills and John Michael Stipe.

'Let's Talk About Sex, Baby': Salt-n-Pepa, Warner/Chappell Music, Inc., 1991. By Herby Azor.

'Life on Mars': David Bowie, Sony/ATV Music Publishing LLC; Peermusic Publishing; Warner/Chappell Music, Inc.; Universal Music Publishing Group, 1971. By David Jones.

'Living Doll': Cliff Richard, Sony/ATV Music Publishing LLX, 1959. By Lionel Bart.

'Monkey Doodle Doo': Mary Eaton, 1913. By Irving Berlin.

'Out of Space': The Prodigy, Warner/Chappell Music, Inc.; Universal Music Publishing Group, 1994. By Liam Howlett.

'Paranoid Android': Radiohead, Warner/Chappell Music, Inc.,

1997. By Colin Greenwood, Edward O'Brien, Jonathan
Greenwood, Philip Selway and Thomas Yorke.
'Psycho Killer': Talking Heads, Warner/Chappell Music, Inc., 1977.
By Chris Frantz, David Byrne and Tina Weymouth
'Take a Walk on the Wild Side': Lou Reed, Sony ATV Music,
1972. By Lewis Reed.

References

1 - The Future is Now

http://www.esquire.com/entertainment/movies/a35559/back-to-the-future-production/

http://www.theverge.com/tldr/2015/10/22/9587400/hendo-hoverboard-back-to-the-future-day

http://physics.stackexchange.com/questions/142732/how-does-hendo-hoverboards-achieve-the-self-propelling-motion-what-is-the-mfa

https://www.kickstarter.com/projects/142464853/hendo-hoverboards-worlds-first-real-hoverboard/description

http://www.solarimpulse.com/leg-8-from-Nagoya-to-Hawaii

http://www.bbc.co.uk/news/science-environment-31772140

http://www.telegraph.co.uk/finance/newsbysector/energy/11129336/Solar-could-beat-coal-as-worlds-top-power-source-by-2050-says-IEA.html

https://education.lego.com/en/about-us/lego-education-worldwide/lego-facts

http://www.financialexpress.com/article/india-news/plastic-paves-way-for-eco-friendly-roads-in-jamshedpur/84203/

http://edition.cnn.com/2014/05/12/tech/solar-powered-roads-coming-highway/

http://www.theguardian.com/world/2015/jul/10/rotterdam-plastic-roads-trial-netherlands

http://www.usatoday.com/story/tech/2016/01/04/ces-2016---meet-worlds-first-smart-bra/78247554/

http://zhor-tech.com/health-and-comfort-improvement/

http://www.nature.com/ncomms/2015/150908/ncomms9094/full/ncomms9094.html

http://www.nature.com/articles/srep12574

http://www.aprilli.com/urban-skyfarm/

http://www.theguardian.com/business/2015/sep/13/
the-innovators-london-air-raid-shelters-sprout-a-growing-concern

http://3dprintingindustry.com/3d-printing-basics-free-beginners-guide/
history/

http://3dprinting.com/bio-printing/

http://www.designboom.com/art/
artist-grows-van-gogh-ear-dna-3d-printer-02-15-2014/

http://3dprint.com/17913/3d-print-ear-clinical-trials/

http://3dprint.com/87806/10-coolest-3d-prints/

http://3dprint.com/12933/3d-printed-castle-complete/

http://3dprint.com/87806/10-coolest-3d-prints/

https://www.theguardian.com/technology/2016/mar/31/chinese-funeral-
home-3d-prints-face-recreation-missing-body-parts-corpses

http://www.livescience.com/54163-nice-trails-3d-printed-mountain-models.
html

http://www.geek.com/news/a-single-3d-printed-burger-currently-costs-
over-300000-to-make-1536823/

https://www.vice.com/en_uk/read/big-brain-connectome-interview

http://www.livescience.com/42561-supercomputer-models-brain-activity.html

http://www.telegraph.co.uk/technology/10567942/Supercomputer-models-
one-second-of-human-brain-activity.html

http://www.smithsonianmag.com/smart-news/
weve-put-worms-mind-lego-robot-body-180953399/?no-ist

http://www.slate.com/articles/health_and_science/science/2016/03/
how_big_is_the_brain_who_knows_even_our_best_efforts_to_
calculate_its_capacity.html

http://www.rd.com/culture/animals-that-live-forever/#

http://www.livescience.com/6967-hang-25-year-wait-immortality.html

2 - Psycho Killer, Qu'est-Ce Que C'est?

https://www.cps.gov.uk/legal/assets/uploads/files/lawyers'%20DNA%20
guide%20KSWilliams%20190208%20(i).pdfhttp://isogg.org/wiki/
Mitochondrial_DNA_tests

http://www.ncbi.nlm.nih.gov/pmc/articles/PMC3192811/

http://www.huffingtonpost.com/2012/12/02/homeland-walden_n_2214685.
html

http://www.huffingtonpost.com/2012/12/02/homeland-brody-
kills_n_2213510.html

http://www.nytimes.com/2008/03/12/business/12heart-web.html?_r=1

http://www.secure-medicine.org

http://www.pacemakerplus.com/pacemaker-pacing-100-what-does-this-mean-what-is-pacing-dependent/

http://www.reuters.com/article/us-cybersecurity-medicaldevices-insight-idUSKCN0IB0DQ20141022

https://www.newscientist.com/article/mg22429942.600-murder-by-hackable-implants-no-longer-a-perfect-crime/?full=true&print=true

http://www.bbc.co.uk/news/technology-33650491

http://www.webmd.com/hypertension-high-blood-pressure/features/health-benefits-of-pets

http://icb.oxfordjournals.org/content/54/2/166.full

http://kreyolicious.com/the-curious-case-of-clairvius-narcisse-and-other-instances-of-haitian-zombies/4005/

http://www.csicop.org/si/show/zombies_and_tetrodotoxin

http://harpers.org/archive/1984/04/the-pharmacology-of-zombies/

http://www.csicop.org/si/show/zombies_and_tetrodotoxin

https://en.wikipedia.org/wiki/The_Serpent_and_the_Rainbow_(book)

http://www.chm.bris.ac.uk/motm/ttx/ttx.htm

http://www.wired.com/2014/02/absurd-creature-of-the-week-jewel-wasp/

http://www.livescience.com/16411-zombies-fact-fiction-infographic.html

http://news.nationalgeographic.com/news/2005/09/0901_050901_wormparasite.html

http://www.sciencedirect.com/science/article/pii/S003193840300163X

http://foreignpolicy.com/2014/05/13/exclusive-the-pentagon-has-a-plan-to-stop-the-zombie-apocalypse-seriously/

http://www.ncbi.nlm.nih.gov/pubmed/26105172

http://www.thebodysoulconnection.com/EducationCenter/fight.html

3 – I'm Gonna Send Him to Outer Space, to Find Another Race

https://www.psychologytoday.com/articles/200303/alien-abductions-the-real-deal

http://www.independent.co.uk/news/science/forget-little-green-men-aliens-will-look-like-humans-says-cambridge-university-evolution-expert-10358164.html

http://www.nasa.gov/ames/kepler/nasas-kepler-discovers-first-earth-size-planet-in-the-habitable-zone-of-another-star

http://www.ijreview.com/2015/08/398055-steven-spielberg-reveals-real-
    life-inspiration-e-t-will-break-heart/
http://news.nationalgeographic.com/
    news/2013/12/131209-curiosity-mars-takeaways-science-life-space/
https://www.sciencedaily.com/releases/2008/01/080125223302.htm
https://www.newscientist.com/article/
    mg18424713-000-pencils-sketch-out-next-electronics-revolution/
http://www.graphene.manchester.ac.uk/explore/what-can-graphene-do/
http://arxiv.org/pdf/1505.04254.pdf
http://www.graphene.manchester.ac.uk
https://www.caltech.edu/news/
    caltech-researchers-find-evidence-real-ninth-planet-49523
http://www.asianscientist.com/2016/01/topnews/
    chinas-wukong-joins-hunt-dark-matter/
http://www.pbs.org/wgbh/nova/next/physics/
    lhc-accidental-rainbow-universe/
http://discovermagazine.com/2013/
    julyaug/23-20-things-you-didnt-know-about-gravity
http://blogs.esa.int/rosetta/2016/02/04/inside-rosettas-comet/
https://www.gov.uk/government/news/countdown-to-comet-touchdown
http://www.independent.co.uk/news/science/philae-lander-bounced-twice-
    on-comet-and-may-still-not-be-stable-rosetta-mission-scientists-warn-
    9857551.html
http://www.space.com/27788-philae-comet-landing-bounce-photos.html
http://www.msn.com/en-au/news/other/
    what-did-we-learn-from-landing-on-a-comet/vi-BBoYXny
https://leahcanscience.com/2014/11/03/light-move-objects/
http://www.bbc.co.uk/news/science-environment-34647921
http://www.bbc.co.uk/news/uk-scotland-17760077

4 - Take a Walk on the Wild Side

http://www.livescience.com/6304-perplexing-panda-pseudo-pregnancy-
    pondered.html
http://journals.plos.org/plosbiology/article?id=10.1371/journal.pbio.0030386
http://www.sciencedaily.com/releases/2015/09/150910185122.htm
http://www.scientificamerican.com/podcast/episode/
    musical-pitch-perception-may-have-long-evolutionary-history/
http://www.popsci.com/during-surgery-cats-prefer-classical-music-ac-dc
http://www.classicfm.com/music-news/latest-news/
    cats-classical-music-relaxing/#BXwaZzEdUss0P085.97

http://www.emedexpert.com/tips/music.shtml

http://news.nationalgeographic.com/2015/03/150313-animals-music-cats-tamarins-psychology-science/

http://www.theguardian.com/lifeandstyle/2013/apr/19/black-sabbath-radio-gardener-chris-beardshaw

http://www.sciencedirect.com/science/article/pii/S0376635713001228

http://www.sciencedirect.com/science/article/pii/S037663579800014X

http://www.ncbi.nlm.nih.gov/pmc/articles/PMC1334394/?page=1

http://www.ncbi.nlm.nih.gov/pmc/articles/PMC1334394/pdf/jeabehav00221-0041.pdf

http://www.ncbi.nlm.nih.gov/pubmed/19533184

https://www.plos.org/wp-content/uploads/2015/11/pone-10-11-Levenson.pdf

http://www.sciencemag.org/news/2015/11/pigeons-spot-cancer-well-human-experts

Cook, R.G., Levison, D.G., Gillett, S.R., Blaisdell, A.P., 'Capacity and limits of associative memory in pigeons', *Psychonomic Bulletin & Review* (2005) 12:350–8.

http://erj.ersjournals.com/content/erj/39/3/511.full.pdf

http://www.pbs.org/wgbh/nova/nature/dogs-sense-of-smell.html

http://edition.cnn.com/2006/HEALTH/02/06/cohen.dogcancerdetect/index.html?iref=newssearch

http://www.bbc.co.uk/news/uk-scotland-34583642

http://www.telegraph.co.uk/news/science/12133470/scientists-test-whether-cats-or-dogs-love-us-more.html

http://www.telegraph.co.uk/news/science/science-news/11681391/Feeling-sad-and-tired-Watch-Grumpy-Cat-say-scientists.html

https://www.theguardian.com/science/2014/jul/23/dogs-jealous-owners-attention-study

http://journals.plos.org/plosone/article?id=10.1371/journal.pone.0094597

http://onlinelibrary.wiley.com/doi/10.1207/S15327078IN0303_6/abstract

http://www.tandfonline.com/doi/full/10.1080/02699930701273716

Harris, C.R., 'A review of sex differences in sexual jealousy, including self-report data, psychophysiological responses, interpersonal violence, and morbid jealousy'. *Personality and Social Psychology Review*, 7 (May 2003), 102–128

http://www.ibtimes.com.au/dogs-hold-grudges-especially-against-people-who-are-mean-their-owners-1452044

http://www.japantimes.co.jp/news/2015/06/12/national/dogs-snub-people-mean-owners-study/#.VqkQQrSDlFK

http://phenomena.nationalgeographic.com/2014/11/07/bats-jam-each-others-sonar/

http://phenomena.nationalgeographic.com/2009/07/17/tiger-moths-jam-the-sonar-of-bats/

http://www.sciencemag.org/news/2014/11/holy-blocked-bat-signal-bats-jam-each-others-calls

http://animals.howstuffworks.com/mammals/13-incredible-bat-facts.htm

https://www.newscientist.com/article/dn4343-fish-farting-may-not-just-be-hot-air/

5 - Paranoid Android

http://theconversation.com/open-letter-we-must-stop-killer-robots-before-they-are-built-44577

https://www.theguardian.com/technology/2015/sep/18/robot-swarms-drone-scientists-hive-mentality

http://www.sciencedirect.com/science/article/pii/S221491471300024X

https://backyardbrains.com/products/roboroach

https://www.youtube.com/watch?v=nnR8fDW3Ilo

https://www.ted.com/talks/a_robot_that_flies_like_a_bird?language=en#t-64752

https://www.sciencedaily.com/releases/2015/03/150331131328.htm

https://www.newscientist.com/article/dn27955-smart-mirror-monitors-your-face-for-telltale-signs-of-disease/

http://www.popsci.com/diy/article/2011-05/2011-invention-awards-picture-health

https://www.theguardian.com/technology/2015/mar/27/google-johnson-and-johnson-artificial-intelligence-surgical-robots

http://www.campaigntoendloneliness.org/threat-to-health/

https://www.theguardian.com/technology/2015/feb/27/robear-bear-shaped-nursing-care-robot

https://www.theguardian.com/lifeandstyle/2016/mar/14/robot-carers-for-elderly-people-are-another-way-of-dying-even-more-miserably

https://www.theguardian.com/technology/2015/dec/31/erica-the-most-beautiful-and-intelligent-android-ever-leads-japans-robot-revolution

http://www.theguardian.com/technology/2015/jun/28/computer-writing-journalism-artificial-intelligence

http://www.guinnessworldrecords.com/world-records/first-android-avatar

http://newsfeed.time.com/2012/03/30/watch-woman-or-machine-sophisticated-japanese-she-bot-blurs-the-line/

https://www.youtube.com/watch?v=nKYTY8P5pZE

http://graphics.bondara.com/Future_sex_report.pdf

http://www.kokoro-dreams.co.jp/english/rt_rent/actroid.html

http://www.news.com.au/news/its-not-about-sex-says-fembot-inventor/
story-fna7dq6e-1111118332678

http://www.livescience.com/1951-forecast-sex-marriage-robots-2050.html

http://www.businessinsider.com/
cyborg-people-who-implanted-tech-2014-8?IR=T#

http://www.irishexaminer.com/business/technology/big-read-when-man-
meets-machine--amal-graafstra-and-his-bio-hacking-body-281773.html

http://discovermagazine.com/2008/jul/23-the-blind-climber-who-sees-
through-his-tongue

https://www.youtube.com/watch?v=WV0bJkk86pw

http://www.wired.com/2015/03/
woman-controls-fighter-jet-sim-using-mind/

6 – Let's Talk About Sex, Baby

Elwin, Verrier, 'The Vagina Dentata Legend', *British Journal of Medical
Psychology*, 19 (1943), pp. 439–53.

Leach, Maria (ed.), *Funk and Wagnalls Standard Dictionary of Folklore Mythology
and Legend*, vol. 2, J–Z (1950), p. 1152.

http://blogs.ucl.ac.uk/researchers-in-museums/2013/03/04/
pulling-teeth-ovarian-teratomas-vagina-dentata/#_ftn1

http://www.medicinenet.com/script/main/art.asp?articlekey=2960

Laqueur, Thomas, *Making Sex: Body and Gender from the Greeks to Freud*
(Harvard University Press, 1992)

http://www.theguardian.com/global-development/2016/jan/13/
menstruation-temples-mosques-india-ban-women

Bhartiya, Aru, 'Menstruation, Religion and Society', *International Journal of
Social Science and Humanity* (November 2013)

http://theappendix.net/blog/2013/6/
this-misterie-of-fucking-a-sex-manual-from-1680

https://www.psychologytoday.com/blog/all-about-sex/201303/hysteria-and-
the-strange-history-vibrators

http://hubpages.com/education/The-Elixir-of-Life-A-Brief-History-of-
Testosterone

http://io9.gizmodo.com/the-man-who-would-give-men-monkey-
testicles-1687010588

Bullock, Shane and Hayes, Majella, *Principles of Pathophysiology* (Pearson, 2012)

http://mashable.com/2015/02/20/history-of-vibrators/#3491t15Rikq7

http://www.nndb.com/people/143/000167639/

http://www.mayoclinic.org/healthy-lifestyle/sexual-health/in-depth/
testosterone-therapy/art-20045728

http://www.nature.com/ijir/journal/v16/n1/full/3901154a.html

https://www.asrm.org/FACTSHEET_Sexual_Dysfunction_and_Infertility/

http://www.telegraph.co.uk/news/uknews/1486054/Raw-oysters-really-are-aphrodisiacs-say-scientists-and-now-is-the-time-to-eat-them.html

http://rewardinthecognitiveniche.us/2012/02/osyters-as-aphrodisiacs-science-based.html

Lippi, D., 'Chocolate and medicine: Dangerous liaisons?' *Nutrition*, 25 (2009), pp. 1100–103.

Liebowitz, Michael, R, *The Chemistry of Love* (Boston: Little, Brown, 1983), pp. 100, 169, 177–8.

Bianchi-Demicheli, F., Sekoranja, L., and Pechère-Bertschi, A., 'Sexuality, heart and chocolate', *Revue Médicale Suisse*, (20 March 2013), 20; 9 (378), pp. 624, 626–9 (http://www.ncbi.nlm.nih.gov/pubmed/23547364)

http://www.ehow.com/about_6077100_horn-rhinoceros-considered-aphrodisiac_.html

Muhammad, I., Zhao, J., Dunbar, D. C., and Khan, I. A., 'Constituents of *Lepidium meyenii* 'maca', *Phytochemistry*, 59 (2002), pp. 105–10.

Zheng, B. L., and others, 'Effect of a lipidic extract from *Lepidium meyenii* on sexual behavior in mice and rats', *Urology*, 55(4), (2000), pp. 598–602.

http://www.sciencedirect.com/science/article/pii/S2050052115301360

https://www.psychologytoday.com/articles/200801/scents-and-sensibility

http://www.ncbi.nlm.nih.gov/pmc/articles/PMC3987372/

http://www.scientificamerican.com/article/human-sexual-responses-boosted-by-bodily-scents/

http://www.cell.com/current-biology/abstract/S0960-9822(14)00327-3

http://www.livescience.com/7023-rules-attraction-game-love.html

http://www.sciencemag.org/content/331/6014/226.abstract

http://www.huffingtonpost.com/2013/04/18/mens-smell-testosterone-attractive-to-women-study_n_3110182.html

http://www.physicsclassroom.com/class/newtlaws/Lesson-4/Newton-s-Third-Law

http://urbanlegends.about.com/library/blsexinspace.htm

http://primary.slate.com/articles/news_and_politics/explainer/2007/02/do_astronauts_have_sex.html

https://www.indiegogo.com/projects/pornhub-space-program-sexploration#/

www.newscientist.com/article/2077807-sound-wave-therapy-is-first-alternative-to-viagra-in-15-years/doi.org/bch9

http://www.healthline.com/health/erection-self-test

http://www.rxlist.com/viagra-drug/clinical-pharmacology.htm

http://www.bbc.co.uk/news/health-29485996

http://www.ectrx.org/forms/ectrxcontentshow.php?year=2008&volume=6
&issue=4&supplement=0&makale_no=0&spage_number=307
&content_type=FULL%20TEXT

http://everything.explained.today/Uterus_transplantation/

https://www.theguardian.com/science/2006/sep/18/medicineandhealth.china

http://www.bbc.co.uk/news/health-31876219

https://www.theguardian.com/education/2014/oct/04/penis-transplants-
anthony-atala-interview

http://video.nationalgeographic.com/video/weirdest-flatworms

http://www.oddee.com/item_97082.aspx

http://news.nationalgeographic.com/news/2001/10/1023_corkscrewduck_2.
html

http://theoatmeal.com/comics/angler

http://whatculture.com/science/10-ridiculously-weird-animal-mating-
rituals?page=10

http://www.neatorama.com/2007/04/30/30-strangest-animal-mating-habits/

http://www.pnas.org/content/98/10/5937.full

Reiss, Diana, *The Dolphin in the Mirror* (Houghton Mifflin Harcourt, 2011)

http://chrisryanphd.com/press-1/

http://www.bbc.co.uk/guides/zg4dwmn

http://www.aces.uiuc.edu/vista/html_pubs/BEEKEEP/CHAPT8/chapt8.
html

https://www.ncbi.nlm.nih.gov/pmc/articles/PMC1892840/

http://www.abc.net.au/news/2016-01-20/new-study-reveals-disease-
fighting-properties-of-bee-semen/7100104

http://rspb.royalsocietypublishing.org/content/282/1818/20151821

http://rspb.royalsocietypublishing.org/content/283/1823/20151785

7 – It's the End of the World ... and I Feel Fine!

http://science.nationalgeographic.com/science/prehistoric-world/
mass-extinction/

http://www.endangeredspeciesinternational.org/overview.html

http://www.bbc.com/news/uk-scotland-glasgow-west-21379024

http://www.space.com/19681-dinosaur-killing-asteroid-chicxulub-crater.
html

http://whc.unesco.org/en/tentativelists/5784/

http://www.livescience.com/50414-chicxulub-crater-drilling.html

https://www.theguardian.com/science/2004/mar/19/taxonomy.science

http://www.theguardian.com/environment/radical-conservation/2015/
oct/20/the-four-horsemen-of-the-sixth-mass-extinction

http://quatr.us/economy/pigs.htm

http://www.smithsonianmag.com/history/how-the-chicken-conquered-the-world-87583657/?no-ist

http://sp.lyellcollection.org/content/395/1/301.short

http://news.nationalgeographic.com/news/2014/05/140529-conservation-science-animals-species-endangered-extinction/

http://www.telegraph.co.uk/news/earth/wildlife/6216775/Chris-Packham-Giant-pandas-should-be-allowed-to-die-out.html

http://www.amphibianark.org/the-crisis/chytrid-fungus/

http://www.adventurescience.org/microplastics.html

http://www.plasticsoupfoundation.org/en/

http://www.ecomare.nl/en/encyclopedia/organizations/environmental-organisations/environmentallobby-ngos/north-sea-foundation/

http://www.beatthemicrobead.org/en/

http://www.cleansea-project.eu/drupal/?q=en/node/19

http://www.plasticoceans.net/crisis/

http://www.nature.com/news/marine-ecology-attack-of-the-blobs-1.9929

http://www.motherjones.com/environment/2014/06/watch-out-summer-swimmers-here-come-jellyfish

http://www.deepseanews.com/2013/07/are-jellyfish-immortal/

http://blogs.ei.columbia.edu/2011/02/26/giant-jellyfish-swarms-%E2%80%93-are-humans-the-cause/

http://news.nationalgeographic.com/2015/07/world-population-expected-to-reach-9-7-billion-by-2050/

http://www.un.org/esa/population/publications/wpm/wpm2001.pdf

http://info.worldbank.org/etools/docs/library/48442/m1s5dixon.pdf

http://www.parliament.uk/about/living-heritage/transformingsociety/electionsvoting/womenvote/overview/thevote/

http://www.diffen.com/difference/Nuclear_Fission_vs_Nuclear_Fusion

http://www.wired.co.uk/news/archive/2015-07/13/mini-ice-age-earth-sunspots

http://www.ksl.com/?nid=367&sid=14906244

'Web-Junkies, China's Addicted Teens', BBC4, Storyville, 2014–15; http://www.bbc.co.uk/programmes/b04hkb5r

http://www.buzzfeed.com/katienotopoulos/50-surprising-facts-about-the-internet#.kqgggKp59z

http://www.bbc.co.uk/news/magazine-32736366

http://www.pbs.org/wnet/nature/rhinoceros-rhino-horn-use-fact-vs-fiction/1178/

http://www.theguardian.com/environment/2011/nov/25/cure-cancer-rhino-horn-vietnam

https://www.savetherhino.org/rhino_info/poaching_statistics

http://rhinorescueproject.org/

http://www.theatlantic.com/business/archive/2013/05/why-does-a-rhino-horn-cost-300-000-because-vietnam-thinks-it-cures-cancer-and-hangovers/275881/

http://www.britannica.com/science/feces

http://www.worldometers.info/world-population/

http://www.bbc.co.uk/news/uk-england-bristol-23333533

http://mmadou.eng.uci.edu/research_biofuelcell.html

http://www.worldcrunch.com/tech-science/this-new-green-fuel-is-yellow-a-car-that-runs-on-urine/urine-power-energy-cars-franco-lisci/c4s14074/

http://www.thewaterpage.com/live-without-water.htm

http://www.watercareer.com.au/archived-news/chip-packet-engineering-to-save-lives

http://blogs.ei.columbia.edu/2011/03/07/the-fog-collectors-harvesting-water-from-thin-air/

http://www.who.int/water_sanitation_health/hygiene/en/

8 - Food, Glorious Food

http://news.bbc.co.uk/1/hi/health/4115506.stm

http://www.telegraph.co.uk/news/worldnews/northamerica/usa/11949643/

http://www.efsa.europa.eu/sites/default/files/scientific_output/files/main_documents/231r.pdf

http://www.naturalnews.com/053163_salmon_Puget_Sound_drug_contamination.html#

https://www.york.ac.uk/research/themes/prozac-and-starlings/

http://science.time.com/2013/12/16/the-triple-whopper-environmental-impact-of-global-meat-production/

http://www.independent.co.uk/environment/climate-change/cow-emissions-more-damaging-to-planet-than-co2-from-cars-427843.html

https://newsinhealth.nih.gov/issue/feb2014/feature1

https://www.newscientist.com/article/mg22530143-900-superbug-risk-from-tonnes-of-antibiotics-fed-to-animals/

https://www.newscientist.com/article/mg22530143-900-superbug-risk-from-tonnes-of-antibiotics-fed-to-animals/

http://www.bbc.co.uk/news/health-34857015

http://journals.plos.org/plosmedicine/article?id=10.1371%2Fjournal.pmed.1001974

http://undergroundhealthreporter.com/
  eating-late-and-obesity/#axzz44Oilbxes
http://www.ncbi.nlm.nih.gov/pubmed/25926415
https://www.sciencedaily.com/releases/2015/12/151223141445.htm
https://www.sciencedaily.com/releases/2011/07/110711151451.htm
http://www.babycentre.co.uk/a25012563/developmental-milestones-taste
http://allaboutbeer.com/article/you%E2%80%99re-better-off-with-beer-
  beer-and-your-health/
http://www.telegraph.co.uk/men/active/mens-health/11383790/Beer-could-
  help-protect-brain-from-Parkinsons-and-Alzheimers.html
http://www.ncbi.nlm.nih.gov/pubmed/25811308
http://www.popsci.com/you-could-eat-fast-food-after-workout-build-muscle
http://gourmethealthychocolates.com/history-of-chocolate/chocolate-facts/
https://www.sciencedaily.com/releases/2014/02/140227092149.htm
http://www.ncbi.nlm.nih.gov/pubmed/20858571
http://www.ncbi.nlm.nih.gov/pubmed/16702322
http://www.ncbi.nlm.nih.gov/pubmed/19735513
http://www.sciencedirect.com/science/article/pii/S0195666316300459
Scholey, A. B., French, S. J., Morris, P. J., Kennedy, D. O., Milne,
  A. L., and Haskell, C. F., 'Consumption of cocoa flavonols results
  in acute improvements in mood and cognitive performance during
  sustained mental effort', *Journal of Psychopharmacology*, 24 (2010),
  pp. 1505–14
http://www.hindawi.com/journals/jchem/2013/289392/
http://www.acs.org/content/acs/en/pressroom/newsreleases/2015/march/
  flavorful-healthful-chocolate-could-be-on-its-way.html
http://www.caffeineinformer.com/top-10-caffeine-health-benefits
http://www.ncbi.nlm.nih.gov/pmc/articles/PMC3554265/
http://www.ncbi.nlm.nih.gov/pubmed/20182054
http://www.medscape.com/viewarticle/823276
http://www.livescience.com/50012-coffee-heart-attack-risk.html
http://onlinelibrary.wiley.com/doi/10.1113/jphysiol.2012.230490/
  abstract;jsessionid=BA969B8A9C9155ADD807D321B55D3303
http://onlinelibrary.wiley.com/doi/10.1113/jphysiol.2012.230490/full
https://thegatewayonline.ca/2015/02/dont-wine-hit-gym-wine-not-gym-
  replacement-u-researchers-say/
http://www.medicaldaily.com/red-wine-burns-fat-and-lowers-blood-
  pressure-plus-5-other-health-benefits-winos-321382
http://www.independent.co.uk/life-style/health-and-families/features/are-
  your-organs-obese-1770238.html
http://diabetes.diabetesjournals.org/content/62/8/2629.full

https://www.jci.org/articles/view/59660
http://www.who.int/nutrition/topics/obesity/en/
http://www.nature.com/nutd/journal/v4/n9/full/nutd201426a.html
http://www.bbc.com/future/story/20141106-the-man-who-makes-diamonds
http://www.livescience.com/2877-chimps-prefer-cooked-food.html
http://www.sci-news.com/othersciences/anthropology/science-homo-pan-
    last-common-ancestor-03220.html
http://www.livescience.com/48078-chimpanzees-learn-behaviors-socially.
    html

9 - We Could Be Heroes (Just for One Day)

https://www.youtube.com/watch?v=Yolum7_0UCA
https://www.youtube.com/watch?v=mdQK_odgedk
https://www.youtube.com/watch?v=LDp1XztObUQ
http://www.news18.com/news/tech/human-carrying-quadcopter-to-flying-
    robot-why-drones-are-not-a-passing-fad-1188655.html
https://www.youtube.com/watch?v=HZRp6iRjnhQ
http://www.isciencetimes.com/articles/4894/20130410/girls-lift-tractor-off-
    dad-oregon-sisters.htm
http://abcnews.go.com/US/superhero-woman-lifts-car-off-dad/
    story?id=16907591
http://www.nbcnews.com/health/
    how-do-people-find-superhuman-strength-lift-cars-921457
https://www.youtube.com/watch?v=-3MxuklTfzk
http://www.darpa.mil/program/restoring-active-memory
http://www.darpa.mil/news-events/2015-09-11a
http://www.sciencealert.com/this-molecule-could-be-the-key-to-
    unlocking-super-memory-in-our-brains
http://news.discovery.com/tech/gear-and-gadgets/novices-download-pilots-
    brainwaves-learn-to-fly-160229.htm
http://www.slate.com/articles/health_and_science/science/2016/03/
    how_big_is_the_brain_who_knows_even_our_best_efforts_to_
    calculate_its_capacity.html
http://www.rsc.org/chemistryworld/2015/10/rolled-polymer-electrode-
    record-brain-activity-without-scarring
http://www.cbsnews.com/news/scientists-study-first-child-with-
    super-memory/
http://www.webmd.com/a-to-z-guides/autoimmune-diseases
http://www.cnet.com/uk/news/scientists-make-quantum-leap-teleport-
    data-farther-than-ever-before/

http://podbay.fm/show/921816230/e/1412607479?autostart=1

https://www.youtube.com/watch?v=pnbJEg9r1o8

https://en.wikipedia.org/wiki/Vulcan_%28Star_Trek%29#Psychology

http://news.berkeley.edu/2011/09/22/brain-movies/

https://www.theguardian.com/science/2014/oct/21/paralysed-darek-fidyka-pioneering-surgery

http://gizmodo.com/these-virtual-reality-experiences-make-you-feel-like-a-1704518885

http://web4.cs.ucl.ac.uk/staff/i.yu/pub/Yu_IEEE_12.pdf

Rizzo, A. A., and Kim, G., 'A SWOT analysis of the field of Virtual Rehabilitation and Therapy', *Presence: Teleoperators and Virtual Environments*, 14 (2005), pp. 1–28.

https://vhil.stanford.edu/mm/2013/rosenberg-plos-virtual-superheroes.pdf